PRACTICE
Workbook

Grade 5

TEXAS HSP Math

Harcourt
SCHOOL PUBLISHERS

Visit *The Learning Site!*
www.harcourtschool.com

Printed in the United States of America

ISBN 13: 978-0-15-356838-1

ISBN 10: 0-15-356838-0

2 3 4 5 6 7 8 9 10 073 16 15 14 13 12 11 10 09 08

Contents

UNIT 6: PERIMETER, AREA, AND VOLUME

Chapter 18: Perimeter

Chapter 19: Area

Chapter 20: Volume

UNIT 7: DATA AND GRAPHING

Chapter 21: Analyze Data

Chapter 22: Display and Interpret Data

UNIT 8: PROBABILITY CHANGE

Chapter 23: Outcomes

Chapter 24: Probability

Spiral Review

Name____ **Lesson 1.1**

Place Value Through Millions

Write the value of the underlined digit.

1. $\underline{1}$89,612,357 ________

2. 51$\underline{2}$,897,934 ________

3. $\underline{8}$3,705 ________

4. 3$\underline{7}$,115,296 ________

5. 2$\underline{5}$4,678,128 ________

6. 63$\underline{1}$,189 ________

7. $\underline{7}$2,334,105 ________

8. $\underline{3}$45,132 ________

9. $\underline{5}$7,912 ________

10. $\underline{1}$2,465,983 ________

11. $\underline{2}$56,245,371 ________

12. 15,2$\underline{7}$9,328 ________

Write each number in two other forms.

13. 647,200

____

____

____

14. 40,000,000 + 20,000 + 1,000 + 80 + 5

____

____

____

What number makes the statement true?

15. 580,000 = 58 × ________

16. 2,760,000 = 276 × ________

Problem Solving and TAKS Prep

17. Fast Fact The diameter of Jupiter is 88,732 miles. How can Michael write the diameter of Jupiter in expanded form?

____

18. Clarrisa learns that the average distance between the Sun and Venus is sixty-seven million miles. How can she write this number in standard form for a poster she is making?

____

19. What is the value of the underlined digit in 7$\underline{2}$9,340,233?

A 20,000

B 20,000

C 2,000,000

D 20,000,000

20. In 358,247,061, which digit is in the hundred thousands place?

F 0

G 2

H 3

J 5

Name____

Lesson 1.2

Understand Billions

Write the value of the underlined digit.

1. 855,283,612,681 ____________
2. 752,801,874,345 ____________
3. 25,908,167,238 ____________
4. 358,354,678,540 ____________
5. 902,851,638,411 ____________
6. 93,668,334,312 ____________

Write the number in two other forms.

7. 50,000,000,000 + 70,000,000 + 8,000,000 + 300,000 + 8,000 + 200 + 5

____

____

8. seventy billion, two hundred seventeen million, five hundred thirty-one

____

____

9. 35,089,207,450

____

____

Problem Solving and TAKS Prep

10. **Algebra** How many dimes equal the same total amount as 1,000,000,000 pennies?

____

11. During a year-long penny drive, a volunteer group collected 10,000,000 pennies. How many stacks of 100 pennies could they make with all of their pennies?

____

12. What is the standard form of fifty-two million, six hundred eight thousand, thirty-nine?

A 52,680,390 **C** 52,608,039

B 52,608,390 **D** 52,068,039

13. In 538,479,247,061, which digit is in the ten billions place?

F 5 **H** 2

G 3 **J** 0

Name____ **Lesson 1.3**

Compare and Order Whole Numbers

Compare. Write <, >, or = for each ◯.

1. 6,574 ◯ 6,547
2. 270,908 ◯ 270,908
3. 8,306,722 ◯ 8,360,272
4. 3,541,320 ◯ 3,541,230
5. 670,980 ◯ 680,790
6. 12,453,671 ◯ 12,543,671

Order from least to greatest.

7. 135,603; 135,306; 135,630

____

8. 89,255; 98,255; 89,525

____

Order from greatest to least.

9. 63,574; 63,547; 63,745

____

10. 5,807,334; 5,708,434; 5,807,433

____

Find the missing digit to make each statement true.

11. 13,625 < 13,6 ____ 7 < 13,649
12. 529,781 > 529,78 ____ > 529,778

Problem Solving and TAKS Prep

USE DATA For 13–14 use the table.

State	Quarters Minted in 2005
California	520,400,000
Minnesota	488,000,000
Oregon	720,200,000
Kansas	563,400,000
West Virginia	721,600,000

13. Which state quarter was minted in the greatest number in 2005?

____

14. Order the number of quarters minted for California, Minnesota, and Oregon from least to greatest.

____

15. Which number is less than 61,534?

A 61,354

B 61,543

C 63,154

D 63,145

16. Which shows the numbers in order from greatest to least?

F 722,319; 722,913; 722,139

G 722,139; 722,319; 722,913

H 722,913; 722,139; 722,319

J 722,913; 722,319; 722,139

Name____________________

Problem Solving Workshop Strategy: Make a Table

Problem Solving Skill Practice

Make a table to solve.

1. The average number of daily passenger flight take-offs and landings at Atlanta Hartsfield is 2,685, at Tokyo Haneda is 781, at London Heathrow is 1,256, and at Chicago O'Hare is 2,664. Order the airports from greatest to least number of take-offs and landings.

____

____

2. Garth made an input/output table using the rule subtract 9. His table shows an input of 28 with an output of 19. What is the output when the input is 17? What is the input when the output is 40?

____

3. An airport parking lot charges $4 per hour or per part of an hour. How much does it charge for 1 hour, 2 hours, 3 hours, and 4 hours?

____

4. A cashier at an airport bookstore earns $7 per hour. How much does the cashier earn in 2 hours, 4 hours, 6 hours, and 8 hours?

____

Mixed Application

USE DATA For 5–7, use the table.

Airline Ticket Sales	
Month	**Tickets sold**
January	201,114
February	176,040
March	100,003
April	187,930
May	142,654
June	194,225
July	127,561

5. Which months had sales under 150,000?

____

6. In August, the airline sold 467 more tickets than it sold in January. How many tickets did it sell in August?

____

7. Find the range, the difference between the greatest number of tickets sold and the least number of tickets sold, during the first 6 months.

____

8. Look back at Exercise 2. Write a similar problem by changing the rule that Garth uses.

____

____

Name____ **Lesson 1.5**

Decimal Place Value

Write the value of the underlined digit.

1. $8.1\underline{3}$ ______
2. $0.\underline{2}6$ ______
3. $9.\underline{4}7$ ______
4. $5.3\underline{6}$ ______
5. $0.\underline{9}2$ ______
6. $0.8\underline{7}$ ______
7. $1\underline{2}.08$ ______
8. $0.\underline{8}1$ ______
9. $1.4\underline{5}$ ______
10. $13.9\underline{4}$ ______

Write each number in two other forms.

11. 5.09 ______ ______

12. 0.84 ______ ______

13. $6 + 0.2 + 0.05$ ______ ______

14. $20 + 0.04$ ______ ______

15. thirty-two and fifty-seven hundredths ______ ______

ALGEBRA Find the missing value(s).

16. $0.38 = (3 \times 0.1) + (\_\_\_\_ \times \_\_\_\_)$

17. $0.92 = (\_\_\_\_ \times \_\_\_\_) + (2 \times 0.01)$

Problem Solving and TAKS Prep

18. Ronald Reagan served as President of the United States for eight years. He was 1.85 meters tall. Write President Reagan's height in expanded form.

______

19. President Richard Nixon was the first president to resign from office. He was 73.5 inches tall. Write the height of President Nixon in word form.

______

______

20. George Washington and Richard Nixon were both one and eighty-seven hundredths meters tall. Write the number one and eighty-seven hundredths in standard form.

______

21. James Madison was 1.63 meters tall. Which shows the expanded form of 1.63?

A $100 + 60 + 3$

B $10 + 6 + 0.3$

C $1 + 0.6 + 0.03$

D $1 + 0.63$

Name____ **Lesson 1.6**

Model Thousandths

Write the decimal shown by the shaded part of each model.

1.

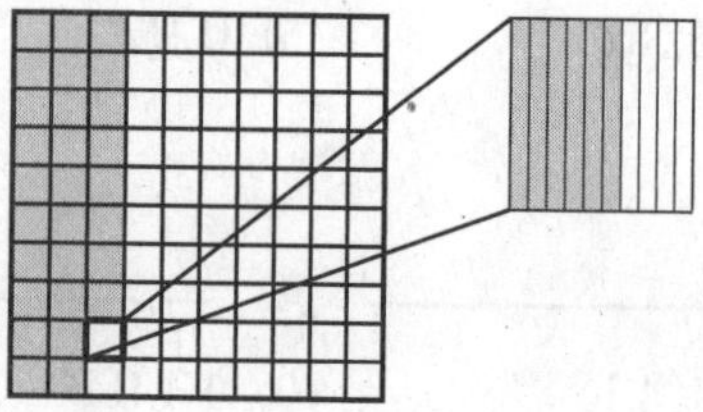

____

2.

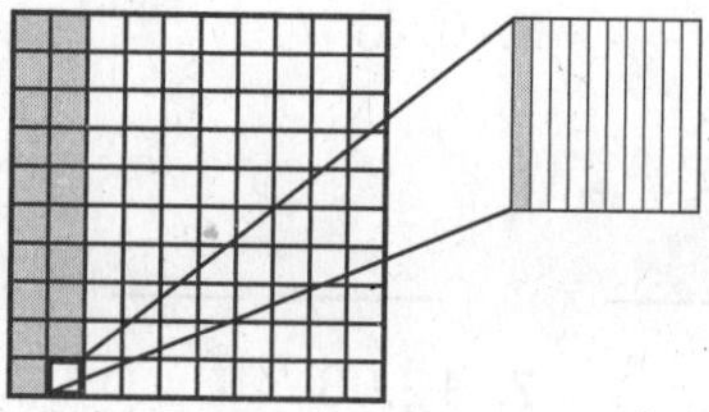

____

3.

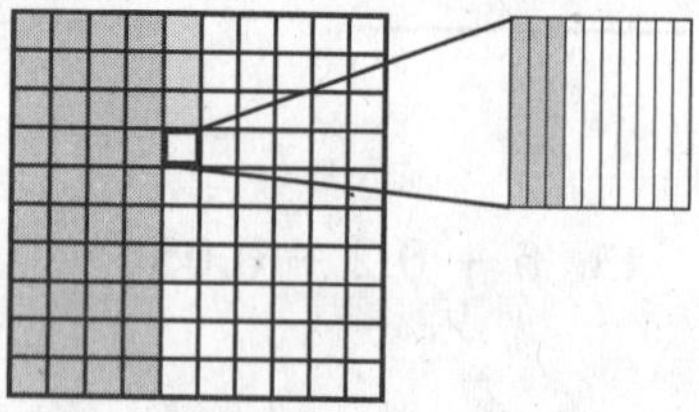

____

4.

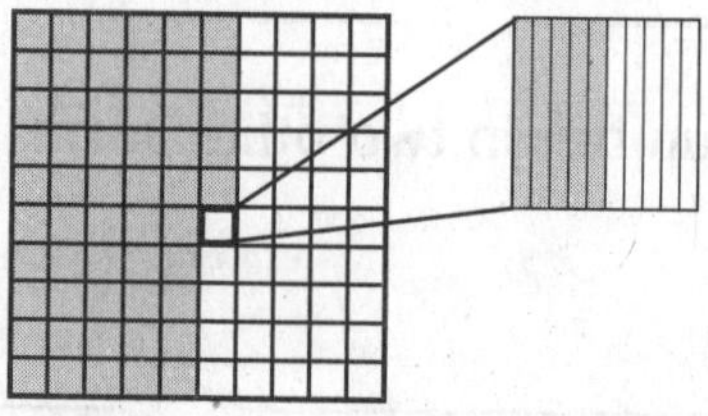

____

Write the value of the underlined digit.

5. 0.72$\underline{5}$ ________

6. 0.0$\underline{1}$8 ________

7. 4.$\underline{0}$93 ________

8. 6.00$\underline{7}$ ________

9. 1.07$\underline{2}$ ________

10. 0.8$\underline{9}$6 ________

11. 0.83$\underline{1}$ ________

12. 2.$\underline{4}$71 ________

13. 3.7$\underline{1}$9 ________

14. 9.$\underline{1}$03 ________

Write each number in two other forms.

15. fifty-four thousandths

____

____

____

16. 0.736

____

____

____

17. 5 + 0.7 + 0.02 + 0.006

____

____

____

18. 3 + 0.2 + 0.009

____

____

____

19. 7.081

____

____

____

20. four and six thousandths

____

____

____

Name____

Equivalent Decimals

Write *equivalent* or *not equivalent* to describe each pair of decimals.

1. 2.26 and 2.260 ____________

2. 8.05 and 8.50 ____________

3. 7.08 and 7.008 ____________

4. 9 and 9.00 ____________

Write an equivalent decimal for each number.

5. 0.9 ____________

6. 1.800 ____________

7. 3.02 ____________

8. 8.640 ____________

9. 0.04 ____________

10. 45.100 ____________

11. 4.60 ____________

12. 2.70 ____________

Write the two decimals that are equivalent.

13. 3.007
3.700
3.7000
____

14. 0.930
0.093
0.93
____

15. 7.60
7.06
7.600
____

16. 3.0540
3.054
3.504
____

Problem Solving and TAKS Prep

17. **Fast Fact** The calliope hummingbird is the smallest bird in North America. It weighs about 2.5 grams and builds a nest about the size of a quarter. Write an equivalent decimal for 2.5.

____

18. The calliope hummingbird is about 0.07 meter long, yet it can fly from northern North America to Mexico for the winter. Write an equivalent decimal for the length of a calliope hummingbird.

____

19. The calliope hummingbird lives in the mountains. It has been seen as high as 335.23 meters above sea level. Write an equivalent decimal for 335.23.

____

20. A banded calliope hummingbird was seen in Idaho and also in Virginia. It had flown more than 2,440.95 miles. Which decimal is equivalent to 2,440.95?

A 2,440.095
B 2,400.905
C 2,440.9500
D 2,440.9595

Name____________________ Lesson 1.8

Compare and Order on a Number Line

Write <, >, or = for each ◯. Use the number line.

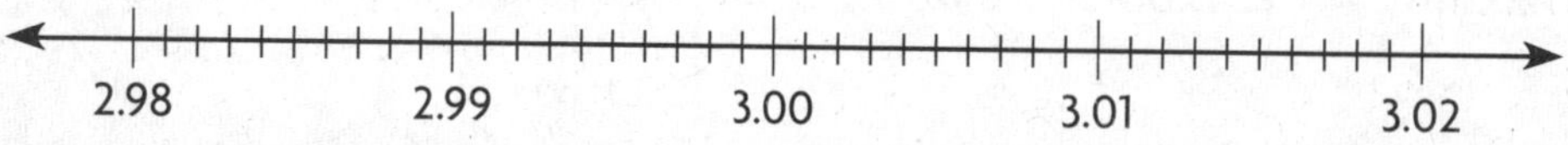

1. 3.007 ◯ 2.993
2. 2.987 ◯ 2.98
3. 3.006 ◯ 3.010
4. 2.90 ◯ 2.901
5. 2.983 ◯ 2.985
6. 3.000 ◯ 3
7. 3.005 ◯ 3.002
8. 3.281 ◯ 3.218
9. 3.015 ◯ 3.009
10. 3.02 ◯ 3.015
11. 2.982 ◯ 3.001
12. 2.98 ◯ 2.980

Order from least to greatest. Use the number line above.

13. 3, 3.02, 2.975, 3.018

____

14. 3.002, 3.025, 2.989, 3.001

____

15. 2.998, 2.898, 2.999, 2.987

____

16. 3.001, 3.010, 3.101, 3.015

____

Problem Solving and TAKS Prep

17. A copper butterfly has a wingspan of 3.285 centimeters. An American snout butterfly has a wingspan of 3.283 centimeters. Which butterfly has the greater wingspan?

____

18. Three butterflies have wingspan lengths of 3.582 cm, 3.285 cm, and 3.283 cm. From least to greatest, what is the order the wingspans?

____

19. Martin wrote the masses of four minerals from his collection below. Which mineral has the greatest mass?

   A 8.553 grams
   B 8.35 grams
   C 8.4 grams
   D 8.299 grams

20. Sam measured the wingspans of four butterflies. Which wingspan is the shortest?

   F 2.01 cm
   G 2.093 cm
   H 2.009 cm
   J 2.10 cm

Name____

Lesson 1.9

Use Place Value to Compare and Order

Compare. Write <, >, or = for each ◯.

1. 0.37 ◯ 0.370 **2.** 3.10 ◯ 3.101 **3.** 0.579 ◯ 0.576 **4.** 7.7 ◯ 7.690

5. 0.812 ◯ 0.821 **6.** 9.810 ◯ 9.809 **7.** 0.955 ◯ 0.95 **8.** 3.281 ◯ 3.218

9. 5.202 ◯ 5.220 **10.** 0.78 ◯ 0.780 **11.** 4.17 ◯ 4.017 **12.** 0.897 ◯ 0.987

Order from least to greatest.

13. 0.301, 0.13, 0.139, 0.5

____

14. 7.203, 7.032, 7, 7.2

____

15. 0.761, 0.67, 0.776, 0.7

____

16. 0.987, 0.978, 0.97, 0.98

____

Problem Solving and TAKS Prep

USE DATA For 17–18, use the table.

17. Which beetle has the shortest length? the longest length?

____

____

18. Another type of beetle is 1.281 cm long. Which beetle has a length less than 1.281 cm?

____

Sizes of Beetles	
Beetle	**Size (in cm)**
Japanese Beetle	1.295
June Bug	2.518
Firefly	1.063

19. Some types of beetles can jump as high as 15 cm. Suppose three beetles jumped 14.03 cm, 14.029 cm, and 14.031 cm. What is the order of the heights the beetles jumped from least to greatest?

____

20. A Japanese Beetle grub may hibernate 29.301 cm underground. Between which two numbers is 29.301?

A 29.103 and 29.300

B 29.21 and 29.3

C 29.3 and 29.31

D 29.31 and 29.32

Name ____________________ **Lesson 2.1**

Round Whole Numbers

Round each number to the place of the underlined digit.

1. 325,689,029 ________
2. 45,673 ________
3. 91,341,281 ________
4. 621,732,193 ________

5. 8,067 ________
6. 42,991,335 ________
7. 182,351,413 ________
8. 539,605,281 ________

9. 999,887,423 ________
10. 76,805,439 ________
11. 518,812,051 ________
12. 657,388,369 ________

Name the place to which each number was rounded.

13. 25,398 to 30,000 ________
14. 828,828 to 830,000 ________
15. 7,234,851 to 7,234,900 ________

16. 612,623 to 600,000 ________
17. 435,299 to 435,000 ________
18. 8,523,194 to 9,000,000 ________

Round 34,251,622 to the place named.

19. millions ________
20. hundred thousands ________
21. thousands ________

Problem Solving and TAKS Prep

22. Fast Fact Wrigley Field in Chicago, Illinois has a seating capacity of 41,118 people. In a newspaper article, that number is rounded to the nearest ten thousand. What number is written in the newspaper article?

____

23. Reasoning The number of seats in Shea Stadium can be rounded to 56,000 when rounded to the nearest thousand. What could be the exact number of seats in Shea Stadium?

____

24. Which number rounded to the nearest million is 45,000,000?

A 43,267,944
B 44,968,722
C 44,322,860
D 44,062,904

25. Which shows 42,167,587 rounded to the nearest million?

F 40,000,000
G 41,000,000
H 42,000,000
J 43,000,000

Name ____________________

Estimate Sums and Differences

Estimate by rounding.

1. 308,222 − 196,231

2. 925,461 − 173,509

3. 19,346 + 25,912

4. 125,689 + 236,817

5. 471,282 − 161,391

6. 305,284 + 2,865,109

7. \$342,199 + \$63,128

8. 78,244 − 23,681

Estimate by using compatible numbers or other methods.

9. 123,636 + 78,239

10. 48,385 + 54,291

11. \$4,471 − \$1,625

12. 69,371 + 73,253

13. 224,119 − 79,388

14. 4,469,235 − 2,328,882

15. 93,215 − 41,284

16. \$246,119 + \$395,228

For 17–20, find the range the estimate will be within.

17. \$3,817 − \$1,428

18. 28,204 + 53,185

19. 35,122 + 61,812

20. 482 + 512

Problem Solving and TAKS Prep

21. Brazil has a population of 186,112,794 people. Argentina has a population of 39,537,943 people. About how many people live in Brazil and Argentina?

22. What if the population of Brazil increased by 4 hundred thousand people, would that change your estimate for problem 22? Explain.

23. Sarah rode her bike 5 days. The longest distance she rode in one day was 6 miles, and the shortest distance she rode is 5 miles. What is a reasonable total number of miles Sarah biked in 5 days?

A Less than 12 mi

B Between 4 mi and 6 mi

C Between 15 mi and 20 mi

D More than 20 mi

24. Use rounding to estimate.

249,118 + 394,417

F 700,000

G 640,000

H 630,000

J 65,000

Name____________________

Add and Subtract Whole Numbers

Estimate. Then find the sum or difference.

1. 6,292 + 7,318

2. 28,434 + 49,617

3. 205,756 − 201,765

4. 529,852 + 476,196

5. 5,071,154 + 483,913

6. 241,933 + 51,209

7. 75,249 − 41,326

8. 1,202,365 − 278,495

9. 4,092,125 + 2,748,810 + 6,421,339

10. 4,687,184 − 1,234,562

11. 542,002 − 319,428

12. 360,219 + 815,364

13. 32,109 + 6,234 + 4,827 ________

14. 3,709,245 − 1,569,267 ________

15. 200,408 − 64,159 ________

ALGEBRA **Find each missing value.**

16. ■ − 1,982 = 8,754

17. 70,380 − ■ = 43,287

18. ■ + 262,305 = 891,411

Problem Solving and TAKS Prep

USE DATA For 19-20, use the table.

19. How many more square miles of surface area does Lake Michigan have than Lake Ontario?

____

20. What is the total surface area of the two lakes with the greatest water surface area?

____

Great Lakes Facts

Lake	Water Surface Area (in sq mi)
Superior	31,700
Michigan	22,300
Ontario	7,340
Erie	9,910
Huron	23,000

21. 328,954 + 683,681 =

A 901,535

B 1,001,535

C 1,012,635

D 1,012,645

22. Over the first weekend in July, the movie theater sold 78,234 tickets. Over the second weekend, the movie theater sold 62,784 tickets. How many more tickets were sold over the first weekend than the second weekend in July?

____

Name________________________________ Lesson 2.4

Choose a Method

Choose a method. Find the sum or difference.

1. $\begin{array}{r} 184{,}621 \\ +\ 715{,}152 \\ \hline \end{array}$

2. $\begin{array}{r} 88{,}632 \\ -\ 72{,}110 \\ \hline \end{array}$

3. $\begin{array}{r} 438{,}197 \\ -\ 428{,}092 \\ \hline \end{array}$

4. $\begin{array}{r} 5{,}001 \\ +\ 3{,}745 \\ \hline \end{array}$

5. $\begin{array}{r} 4{,}207 \\ -\ 459 \\ \hline \end{array}$

6. $\begin{array}{r} 60{,}000 \\ +\ 5{,}400 \\ \hline \end{array}$

7. $\begin{array}{r} 85{,}000 \\ -\ 4{,}992 \\ \hline \end{array}$

8. $\begin{array}{r} 306{,}781 \\ -\ 104{,}825 \\ \hline \end{array}$

9. 39,000 − 15,000 ________

10. 17,189 − 7,000 ________

11. 4,089 + 12,000 ________

12. 28,144 + 36,291 ________

13. 150,000 − 102,000 ________

14. 319,128 + 495,000 ________

ALGEBRA Find each missing value.

15. ■ − 8,846 = 1,400

16. ■ + 1,296 = 1,735

17. 7,463 − ■ = 1,511

Problem Solving and TAKS Prep

USE DATA For 18–19, use the table.

18. Justin mixed 1 cup of peanuts with 1 cup of cashews. How many Calories were in the mixture?

____

19. How many fewer Calories are in 1 cup of pecans than in 1 cup of cashews?

____

Popular Nuts

Nut (1 cup)	Calories
Black walnuts	642
Cashews	787
Peanuts	822
Pecans	721

20. Mrs. Scott paid $14,502 for a used car. This price was $8,848 less than what she paid for a new car 4 years ago. How much did she pay 4 years ago?

A $5,654

B $6,346

C $22,340

D $23,350

21. Mr. Roy earned $28,436 last year. This was $1,080 more than he earned the year before. How much did he earn the year before?

F $27,356

G $27,456

H $29,416

J $29,516

Name________________________________ Lesson 2.5

Problem Solving Workshop Strategy: Look for a Pattern

Problem Solving Skill Practice

Find a pattern to solve the problem.

1. Anna paid $515 monthly rent the first year, $540 monthly rent the second year, $565 monthly rent the third year, and $590 monthly rent the fourth year. If this pattern continues, how much monthly rent will Anna pay in the sixth year?

____

2. On the Coastal Trail, hikers walked 28 miles on Monday, 27 miles on Tuesday, 25 miles on Wednesday, and 22 miles on Thursday. How many miles did the hikers walk on Sunday?

____

3. What are the next three numbers in the pattern?

1, 121, 12321, 1234321, . . .

____

4. A certain redwood tree was 175 ft tall in 2000, 179 ft in 2001, 183 ft in 2002, and 187 ft in 2003. How tall will it be in 2009?

____

Mixed Application

USE DATA For 5–6, use the table.

5. Predict the membership of the Friendship Club for the year 2010.

____

6. In 2006, the membership was twice what it was in 2002. What was the membership in 2006?

____

Friendship Club Membership

Year	Membership
2000	6
2001	12
2002	18
2003	24
2004	30

7. The tallest known Redwood in Redwood National Park was 372 feet tall before it fell in 1991. Yosemite Falls is a little more than 6.5 times taller than that. About how tall is Yosemite Falls?

____

8. Jana spent $182 on a winter coat, $19 on a hat, $8 on a scarf, $6 on gloves, and $21 on boots. How much did Jana spend on her winter apparel?

____

Name ____________________

Round Decimals

Round each number to the place of the underlined digit.

1. 54.2$\underline{4}$7 ________
2. 0.$\underline{1}$09 ________
3. 7.$\underline{0}$44 ________
4. 1$\underline{2}$.581 ________
5. 0.0$\underline{0}$3 ________
6. 4.$\underline{6}$59 ________
7. 8.$\underline{9}$06 ________
8. 0.$\underline{9}$81 ________
9. 23.1$\underline{3}$2 ________
10. 3.4$\underline{9}$6 ________

Round to the nearest tenth of a dollar and to the nearest dollar.

11. $0.78 ________ ________
12. $1.24 ________ ________
13. $0.11 ________ ________
14. $25.54 ________ ________
15. $13.49 ________ ________
16. $0.92 ________ ________
17. $2.95 ________ ________
18. $6.33 ________ ________
19. $20.02 ________ ________
20. $19.59 ________ ________

Problem Solving and TAKS Prep

USE DATA For 21–22, use the graph.

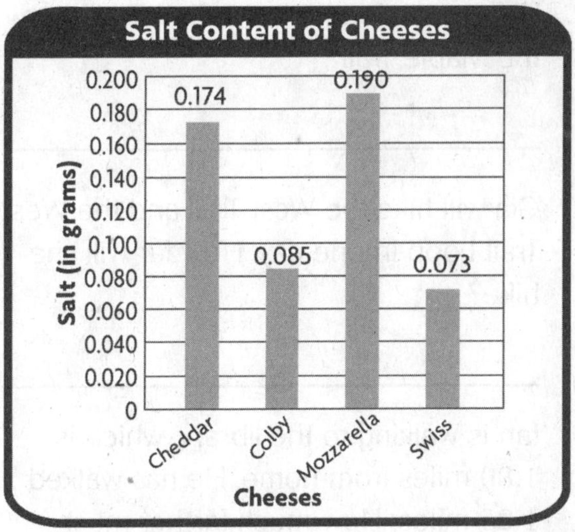

21. Round the salt content of mozzarella cheese to the nearest tenth of a gram.

____

22. Which cheese has a salt content of 0.17 when rounded to the nearest hundredth of a gram?

____

23. Greta rounded 6.488 pounds to 6.49 pounds. To which place did she round?

   A Ones
   B Tenths
   C Hundredths
   D Thousandths

24. Neil rounded 9.135 pounds to 9.1 pounds. To which place did he round?

   F Ones
   G Tenths
   H Hundredths
   J Thousandths

Name____________________

Add and Subtract Decimals

Find the sum or difference.

1. 2.5 + 4.6

2. 4.9 − 3.8

3. 8.1 − 2.4

4. 1.17 + 6.25

5. 10.31 + 7.27

6. 0.22 + 0.85

7. 2.7 − 1.9

8. 1.18 + 0.92

9. 5.47 − 4.38

10. 6.14 − 5.24

11. 7.25 + 1.43

12. 18.2 − 17.7

13. 1.55 − 1.48

14. 26.7 + 14.9

15. 5.26 − 4.33

16. 9.91 + 0.31

17. 1.32 + 4.81 + 7.27

18. 4.58 − 3.67

19. 4.6 + 1.95 + 0.3 ____________

20. 1.28 − 0.44 ____________

21. 0.86 + 0.24 ____________

22. 6.05 − 5.92 ____________

Problem Solving and TAKS Prep

USE DATA For 23–24, use the table.

Lost Maples State Park Trails

Trail Name	Trail Length (miles)
East Trail	4.20
West Trail	3.40
West Trail Loop	1.80
Maple Trail	0.27

23. How much longer is the East Trail than the Maple Trail?

____

24. Gia will hike the West Trail and the West Trail Loop in one day. How far will she hike?

____

25. Ian is walking to the library, which is 1.80 miles from home. He has walked 1.25 miles. How much farther must Ian walk?

A 0.65 mile

B 0.6 mile

C 0.55 mile

D 0.5 mile

26. Eva bicycled to a store, which is 3.44 miles from her home. Then she bicycled to a friend's house 1.75 miles from the store. How far did she pedal?

F 4.19 miles

G 4.29 miles

H 5.19 miles

J 5.29 miles

Name____________________

Lesson 3.3

Estimate Decimal Sums and Differences

Estimate by rounding.

1. $6.7 - 4.8$

2. $10.238 + 7.842$

3. $2.11 + 0.96$

4. $14.54 - 7.35$

5. $9.786 - 8.914$

6. $3.28 + 3.65$

7. $6.14 + 4.79$ ______

8. $12.3 - 2.85$ ______

9. $1.184 + 1.295$ ______

10. $8.72 - 5.43$ ______

Estimate by using compatible numbers.

11. $9.276 + 4.758$

12. $0.63 + 0.31$

13. $10.82 - 5.78$

14. $1.53 - 0.15$

15. $5.34 + 2.68$

16. $4.29 - 3.334$

17. $0.219 + 0.183$ ______

18. $3.64 - 0.58$ ______

19. $14.12 + 5.85$ ______

20. $15.41 - 4.96$ ______

Problem Solving and TAKS Prep

USE DATA For 21-22, use the table.

Hiking Gear	
Item	**Cost**
Hiking boots	$84.87
Compass	$19.95
Insect repellent	$5.25

21. Andrea has $150 to purchase the hiking gear. About how much change should Andrea receive if she buys all three items?

______

22. Estimate how much more the hiking boots cost than the compass.

______

23. Elise has $300 to buy school supplies for $84.90 and a winter coat for $175.29. About how much money will Elise have left?

A Less than $35

B Between $35 and $45

C Between $45 and $55

D More than $55

24. Mario biked 4 days last week. The farthest he biked in one day was 8.4 miles. The shortest was 6.8 miles. Which is a reasonable total number of miles Mario biked during the 4 days?

F Less than 12 mi

G Between 12 and 24 mi

H Between 24 mi and 36 mi

J More than 36 mi

Name______________________________ Lesson 3.4

Add and Subtract Decimals through Thousandths

Estimate. Then find the sum or difference.

1. $0.67 + 0.5$

2. $11.7 - 3.04$

3. $2.96 + 0.045 + 8.62$

4. $9.8 - 0.82$

5. $32.44 - 4.78$

6. $0.45 + 0.071$

7. $0.868 - 0.23$

8. $18.394 + 15.602$

9. $9.46 - 0.5$

10. $25.73 + 15.48$

11. $8 - 4.091$

12. $0.12 + 1.095$

13. $1.304 - 1.239$

14. $0.49 + 0.561 + 2.7$

15. $24.006 - 2.73$

16. $8.18 + 0.517 + 1.304$

17. $0.1 - 0.025$

18. $0.775 + 5.31 + 3.016$

19. $0.003 + 1 + 9.44$

Problem Solving and TAKS Prep

20. Until the 2002 Olympics, the record luge speed was 85.38 miles per hour. Tony Benshoof broke that record with a speed of 86.6 miles per hour. By how much did Tony Benshoof exceed the record?

____

21. Beth and her grandmother paid $23.00 for tickets to a play. An adult ticket costs $6.50 more than a child's ticket. What was the cost of Beth's ticket?

____

22. Lynne buys a meal and milk at the school cafeteria. If she pays with a $5 bill, how much change should Lynne receive?

A $1.06

B $1.55

C $2.96

D $3.94

School Cafeteria	
meal	$3.45
fruit	$0.80
milk	$0.49

23. Tim buys a daily planner and pen at the school store. He pays with a $20 bill. How much change should Tim receive?

F $9.76

G $9.86

H $10.24

J $16.74

School Store	
notebook	$4.55
12 pencils	$2.14
1 pen	$1.29
daily planner	$8.95

Name________________________________

Choose a Method

Choose a method. Find the sum or difference.

1. $\begin{array}{r} 8.24 \\ +\ 0.673 \\ \hline \end{array}$

2. $\begin{array}{r} 7.89 \\ -\ 3.21 \\ \hline \end{array}$

3. $\begin{array}{r} 41.621 \\ -\ 38.94 \\ \hline \end{array}$

4. $\begin{array}{r} \$12.56 \\ +\ \$25.72 \\ \hline \end{array}$

5. $\begin{array}{r} 3.1 \\ 4.75 \\ +\ 2.9 \\ \hline \end{array}$

6. $\begin{array}{r} \$14.27 \\ +\ \$\ 8.49 \\ \hline \end{array}$

7. $\begin{array}{r} 4.803 \\ -\ 2.77 \\ \hline \end{array}$

8. $\begin{array}{r} \$21.40 \\ -\ \$20.10 \\ \hline \end{array}$

9. $\begin{array}{r} \$13.60 \\ -\ \$11.32 \\ \hline \end{array}$

10. $\begin{array}{r} 6.33 \\ 4.095 \\ +\ 1.708 \\ \hline \end{array}$

11. $\begin{array}{r} 0.501 \\ +\ 6.79 \\ \hline \end{array}$

12. $\begin{array}{r} 2.9 \\ -\ 1.5 \\ \hline \end{array}$

13. $\begin{array}{r} 3.37 \\ +\ 6.73 \\ \hline \end{array}$

14. $\begin{array}{r} \$57.19 \\ +\ \$\ 2.68 \\ \hline \end{array}$

15. $\begin{array}{r} 1.005 \\ -\ 0.07 \\ \hline \end{array}$

16. 2.4 + 3.75 + 1.8 ____________

17. 0.85 − 0.798 ____________

18. $1.95 + $7.65 ____________

19. 5.4 − 0.54 ____________

Problem Solving and TAKS Prep

USE DATA For 20–21, use the table.

Women's Long Jump Records		
Name	**Year**	**Distance (in meters)**
Galina Chistyakova	1988	7.52
Jackie Joyner-Kersee	1994	7.49
Heike Dreschler	1992	7.48
Anis oara Stanciu	1983	7.43
Tatyana Kotova	2002	7.42
Yelena Belevskaya	1987	7.39

20. How much farther did Chistyakova jump in 1988 than Joyner-Kersee in 1994?

____

21. What is the difference in jump distances from the earliest listed date to the latest listed date?

____

22. Lydia has 3 dimes, a quarter, a dollar, and 2 nickels. How much money does Lydia have? Show your work.

____

23. Dylan has 2 dollars, 3 quarters, 4 dimes, and a nickel. How much money does Dylan have? Show your work

____

Name____ **Lesson 3.6**

Problem Solving Workshop Skill: Evaluate Answers for Reasonableness

Problem Solving Skill Practice

Evaluate answers for reasonableness.

1. Owen pays $2.50 per day for a round trip ticket. The Trinity Railway Express offers a discount for a monthly pass. The railway expects most passengers to use the pass during weekdays only.

Lora says the monthly pass should cost about $70. Owen says it should only cost about $40. Who has the more reasonable answer? Explain.

____

____

____

____

2. Celia walks 1.5 miles to and from school every day. Based on her weight, she burns about 60 Calories per mile. The school year is 180 days. Celia wants to know how many Calories she burns.

Dave says she burns 4.5 Calories. Danielle says she burns 16,200 Calories. Who is right? Explain.

____

____

____

____

Mixed Applications

3. Tom has 21 flowering plants in white, pink, and lavender flowers. He has 2 more pink than lavender plants. What is the greatest possible number of white flowering plants that he has?

____

4. At noon, the temperature was 58°F. In the next hour, the temperature rose 2°. The hour after that, it rose 4°. During the following hour the temperature rose 6°, and the hour after that, it rose 8°. If the pattern continues, what is the temperature at 6:00 P.M.?

____

5. How many legs do 9 chickens and 23 cows have altogether?

____

6. **Pose a Problem** Look at Exercise 4. Change the beginning temperature in the problem. Then solve it.

____

Mental Math: Patterns in Multiples

Find the product.

1. 9×300 ______
2. 3×100 ______
3. 60×5 ______
4. $5 \times 7{,}000$ ______
5. 60×40 ______
6. 70×200 ______
7. 20×900 ______
8. 40×30 ______
9. 500×30 ______
10. $6{,}000 \times 8$ ______
11. 40×900 ______
12. 7×200 ______
13. 600×60 ______
14. 100×60 ______
15. 60×500 ______

ALGEBRA Find the missing number.

16. $12 \times 400 =$ ______
17. ______ $\times 20 = 1{,}000$
18. $600 \times$ ______ $= 1{,}200$
19. ______ $\times 100 = 11{,}000$
20. $30 \times 500 =$ ______
21. $400 \times$ ______ $= 4{,}000$
22. $50 \times 200 =$ ______
23. $40 \times$ ______ $= 2{,}000$
24. ______ $\times 80 = 4{,}000$

Problem Solving and TAKS Prep

25. One colony of macaroni penguins has about 8,000 nests. If three penguins occupy each nest, how many penguins are there in all?

______

26. Each pair of macaroni penguins lays 2 eggs. How many eggs do 12,000,000 pairs of penguins lay?

______

27. Tickets to a baseball game cost $90 each. How much money will be made in ticket sales if 5,000 tickets are sold?
   - **A** $45,000
   - **B** $450,000
   - **C** $4,500,000
   - **D** $45,000,000

28. A sedan at a car dealership sells for $20,000. How much money will 200 of them bring in?
   - **F** $40,000
   - **G** $400,000
   - **H** $4,000,000
   - **J** $40,000,000

Name____________________

Estimate Products

Estimate the product.

1. 65 × 22 **2.** 27 × $34.12 **3.** 738 × 59 **4.** 195 × 231 **5.** 8,130 × 77

6. 81 × $79.51 **7.** 641 × 312 **8.** 555 × 470 **9.** 4,096 × 12 **10.** 42 × 51,912

11. 199 × 249 **12.** 467 × 124 **13.** 88 × 27 **14.** 411 × 6,725 **15.** 6,371 × 52

16. 33 × 180 **17.** 894 × 605 **18.** 50,720 × 79 **19.** $40.16 × 49 **20.** 76 × 45,118

Problem Solving and TAKS Prep

USE DATA For 21–22, use the table.

Green Park Expenses	
Tree	**Cost**
Southern Magnolia	$11
Redbud	$9
Bigtooth Maple	$18

21. The Municipal Park Committee has budgeted $500 for 35 bigtooth maple trees for Green Park. Estimate to find whether this is enough to buy the trees.

____

22. The park committee also wants to purchase 24 Southern magnolias. Estimate to find whether $300 is enough to purchase them.

____

23. Which would give the best estimate for 48 × 54,090?

A 40 × 50,000

B 40 × 60,000

C 50 × 50,000

D 50 × 60,000

24. Which would give the best estimate for 108 × 276?

F 100 × 200

G 100 × 300

H 200 × 200

J 200 × 300

Name ______________________________

Multiply by 1-Digit Numbers

Estimate. Then find the product.

1. 48×2 ________
2. 317×9 ________
3. 105×3 ________
4. 477×7 ________
5. 729×8 ________
6. 6×802 ________
7. 4×426 ________
8. 339×5 ________
9. 307×4 ________
10. 9×218 ________
11. 531×2 ________
12. 372×8 ________
13. 47×6
14. 26×6
15. 207×3
16. 783×9
17. 428×5
18. 339×7
19. 518×5
20. 309×8
21. 801×3
22. 937×6

Problem Solving and TAKS Prep

USE DATA For 23–24, use the table.

Round Trip Airfares from Dallas, TX

Destination	Cost in Dollars
Honolulu, HI	$619
London, England	$770
Vancouver, WA	$358

23. How much would it cost a family of 6 to fly roundtrip from Dallas to Vancouver?

____

24. How much more does it cost 2 people to fly roundtrip from Dallas to London than to fly from Dallas to Honolulu?

____

25. Which expression has the same value as $8 \times (800 + 70 + 3)$?
   - **A** $8 \times (800{,}703)$
   - **B** $64 + 56 + 24$
   - **C** $6400 + 70 + 3$
   - **D** $6400 + 560 + 24$

26. New windows cost $425 each. What is the total cost for 9 windows?
   - **F** $3,725
   - **G** $3,825
   - **H** $4,725
   - **J** $4,825

Name________________________________ Lesson 4.4

Multiply by 2-Digit Numbers

Estimate. Then find the product.

1. 34×28
2. 451×61
3. 709×53
4. 622×34
5. 970×17
6. $\$229 \times 77$
7. 907×83
8. 385×48
9. 172×91
10. 409×67
11. 219×84
12. 727×33
13. $\$948 \times 58$
14. 122×62
15. 893×312

ALGEBRA Find the missing digit. Explain your solution.

16. 45___ × 47 = 21,432
17. 149 × 9___ = 14,006
18. 52___ × 36 = 19,008

Problem Solving and TAKS Prep

19. Abby wants to cycle 25 miles each day for one full year, or 365 days. How many miles is Abby planning to cycle in all?

20. Randi participated in a Bike-a-Thon. Twenty-three family members donated $12 for each mile she rode. If Randi rode 38 miles, how much did she collect?

21. How much does a store make if it sells 39 computers at $936 each?

A $11,232

B $11,304

C $36,364

D $36,504

22. Kane is training for a cycling event on a track in which one lap is 880 yards. How far has Kane ridden in 12 laps?

F 2,496 yards

G 2,640 yards

H 10,460 yards

J 10,560 yards

Name________________________________ **Lesson 4.5**

Practice Multiplication

Estimate. Then find the product.

1. 3 × 782 **2.** 913 × 7 **3.** 205 × 4 **4.** 5 × 839 **5.** 970 × 6

6. 89 × 306 **7.** 914 × 93 **8.** 9 × 391 **9.** 60 × 224 **10.** 406 × 23

11. 79 × 528 **12.** 619 × 44 **13.** 229 × 36 **14.** 54 × 379 **15.** 672 × 8

Problem Solving and TAKS Prep

USE DATA For 16–17, use the table.

How much fruit zoo animals eat per week	
Fruit	**Approximate weight**
apple	78 pounds
banana	46 pounds
grapes	18 pounds

16. Grapes cost $4 per pound. How much does one week's worth of grapes cost at the zoo?

____

17. How much more do 2 weeks worth of apples weigh than 2 weeks worth of bananas?

____

18. A theme park sells one-day family passes for $98. How much did 687 families pay for passes in one day?

A $66,226
B $66,326
C $67,226
D $67,326

19. Admission to a zoo is $17 per car. How much money did the zoo receive from 2,631 cars in one week?

F $40,527
G $40,737
H $44,727
J $47,054

Name______________________________ Lesson 4.6

Choose a Method

Find the product. Choose mental math, paper and pencil, or a calculator.

1. 500×12 **2.** 375×218 **3.** $40 \times 5{,}000$ **4.** 112×83 **5.** 13×600

______ ______ ______ ______ ______

6. 820×10 **7.** $5{,}129 \times 18$ **8.** 452×726 **9.** 304×21 **10.** $1{,}200 \times 12$

11. 400×320 **12.** 785×122 **13.** $93 \times 11 \times 34$ **14.** $40 \times 10 \times 200$

______ ______ ______ ______

Problem Solving and TAKS Prep

USE DATA For 15–16, use the table.

Animal Sleep	
Animal	**Time (hours per day)**
Tiger	16
Pig	9
Cow	4

15. How many hours does a tiger sleep in one year?

____

16. In one year, how much longer does a pig sleep than a cow?

____

17. An African elephant may weigh 185 pounds at birth. At maturity, it can weigh about 32 times that. What does the elephant weigh at maturity?

A 3,710 pounds

B 4,920 pounds

C 5,920 pounds

D 6,910 pounds

18. A giraffe can weigh 145 pounds at birth. At maturity, it can weigh about 18 times that. What does a giraffe weigh at maturity?

F 1,075 pounds

G 1,305 pounds

H 2,380 pounds

J 2,610 pounds

Name________________________________ Lesson 4.7

Problem Solving Workshop Strategy: Write an Equation

Problem Solving Strategy Practice

Write and solve an equation for each problem.

1. Mark Kelly spent 12 days in space. This is 8 times the amount of time a new astronaut is scheduled to spend in space. How long, in hours, will a new astronaut spend in space?

____

2. Geri has worked at ProWorks Company 3 times as many years as Miguel. Geri has worked at ProWorks 24 years. How long has Miguel worked at ProWorks?

____

3. A certain orchestra has 6 times as many violins as cellos. If there are 54 violins, how many cellos are there?

____

4. Hannah spent 6 hours practicing her violin last week. This was 4 times the amount of time Derek spent. How long, in minutes, did Derek practice?

____

Mixed Applications

USE DATA For 5–6, use the table.

5. Yelena Vladimirovna Kondakova, a Russian cosmonaut, spent 32 fewer days in space than Susan Helms. How many days did Yelena spend in space?

____

Women in Space	
Name	**Days in Space**
Shannon Lucid	223
Susan Helms	210
Peggy Whitson	184

6. How many more days did Shannon Lucid spend in space than Peggy Whitson?

____

7. Helena has 9 times as many CDs as Vinnie. If Helena has 144 CDs, how many does Vinnie have?

____

8. Portia buys a book for $14.85. She pays with a $10 bill and a $5 bill. What is the fewest number of coins she gets back?

____

Name________________________________

Estimate with 1-Digit Divisors

Estimate the quotient.

1. $2\overline{)239}$ **2.** $6\overline{)534}$ **3.** $5\overline{)385}$ **4.** $8\overline{)256}$

5. $224 \div 7$ **6.** $328 \div 4$ **7.** $513 \div 9$ **8.** $294 \div 6$

9. $7\overline{)553}$ **10.** $5\overline{)240}$ **11.** $9\overline{)756}$ **12.** $6\overline{)492}$

13. $501 \div 7$ **14.** $368 \div 6$ **15.** $572 \div 8$ **16.** $794 \div 2$

Problem Solving and TAKS Prep

17. A shipment of cross-country bikes weighs 208 pounds. The shipment included 8 bicycles. About how much did each cross-country bike weigh?

18. Another shipment of bicycles was sent weighing 276 pounds. This shipment included 6 mountain bikes. About how much did each mountain bike weigh?

19. Mr. Jones drove 571 mi in 4 days. Which is the best estimate of how far Mr. Jones drove on the first day?

A 162 mi
B 140 mi
C 115 mi
D 96 mi

20. John traveled 885 mi in 3 days. Which is the best estimate of how far John drove on the first day?

F 190 mi
G 268 mi
H 300 mi
J 250 mi

Name________________________________

Divide by 1-Digit Divisors

Name the position of the first digit of the quotient. Then find the first digit.

1. $4\overline{)392}$ **2.** $7\overline{)868}$ **3.** $5\overline{)730}$ **4.** $6\overline{)468}$ **5.** $9\overline{)783}$

______ ______ ______ ______ ______

6. $3\overline{)807}$ **7.** $8\overline{)592}$ **8.** $7\overline{)532}$ **9.** $9\overline{)387}$ **10.** $8\overline{)904}$

______ ______ ______ ______ ______

Divide.

11. $2\overline{)684}$ **12.** $5\overline{)745}$ **13.** $7\overline{)679}$ **14.** $4\overline{)276}$ **15.** $8\overline{)372}$

16. $854 \div 6$ **17.** $260 \div 3$ **18.** $657 \div 9$ **19.** $962 \div 7$ **20.** $519 \div 4$

______ ______ ______ ______ ______

Problem Solving and TAKS Prep

21. 139 students are going to a museum by van. Each van can hold 8 students. How many full vans are needed? How many students are riding in the van that is not full?

____

22. There are 139 students at the museum. Each adult has 9 students in their group. How many adults will have a full group? How many students will not be in a 9-student group?

____

23. One case can hold 9 boxes of cereal. How many cases are needed to hold 162 boxes of cereal?

A 1,377
B 18
C 17
D 9

24. A fifth-grade class made 436 cookies. The class put 6 cookies in each bag. How many cookies remained?

F 72 r4
G 2,616
H 4
J 72

Name ____________________ Lesson 5.3

Zeros in Division

Divide.

1. $8\overline{)928}$

2. $3\overline{)738}$

3. $6\overline{)521}$

4. $7\overline{)933}$

5. $4\overline{)743}$

6. $5\overline{)597}$

7. $6\overline{)648}$

8. $4\overline{)857}$

9. $3\overline{)980}$

10. $7\overline{)829}$

11. $619 \div 9$ ______

12. $724 \div 2$ ______

13. $965 \div 8$ ______

14. $858 \div 6$ ______

15. $531 \div 4$ ______

Problem Solving and TAKS Prep

16. Each pack of marigold flowers can hold 6 marigolds. There are 458 marigolds. How many full packs of marigolds are there? How many more marigolds are needed to fill a 6-pack of marigolds?

______

17. Each pack of tulips can hold 9 tulips. There are 956 tulips to be packed. How many tulips will be left? How many more tulips are needed to fill a 9-pack container of tulips?

______

18. The population of the world in July 2006 was about 6,628,506,453. What is the value of the digit 2 in that number?

______

19. A pet store sells dog bones in packages of 6. How many packages can they make from 762 dog bones?

A 127

B 4,572

C 6

D 172

Name ______________________ **Lesson 5.4**

Practice Division

Divide.

1. $6\overline{)456}$ **2.** $8\overline{)376}$ **3.** $9\overline{)778}$ **4.** $5\overline{)726}$ **5.** $7\overline{)919}$

6. $386 \div 4$ **7.** $323 \div 5$ **8.** $846 \div 6$ **9.** $495 \div 4$ **10.** $629 \div 3$

Find the average for each set of numbers.

11. 37, 42, 46, 31, 44 **12.** 168, 206, 193, 225 **13.** 92, 134, 97, 114, 128

Problem Solving and TAKS Prep

14. Julia can make a paper crane in 7 minutes. She spent 882 minutes making paper cranes for a party. How many paper cranes did Julia make?

15. Nathan spent 696 minutes making paper origami boxes. He can make a paper box in 6 minutes. How many origami boxes did Nathan make?

16. Sean has 150 coins in his collection. He has 85 more than Jane and 70 fewer coins than Steven. How many coins does Steven have in his collection?

17. A school cafeteria used 234 pieces of bread yesterday equaling 9 full loaves. How many pieces of bread are in one loaf?

A 24

B 25

C 26

D 27

Name________________________________

Problem Solving Workshop Skill: Interpret the Remainder

Tell how you would interpret the remainder. Then give the answer.

1. A total of 110 fifth graders are going on a field trip to a museum. Vans will be used for transportation. Each van holds 8 students. How many vans will be needed for the trip?

____

____

2. The Jones family is planning a hiking trip in the mountains. The Joneses want to hike 9 miles each day. How many days will it take for the Jones family to hike 114 miles? How many miles will they hike on the last day?

____

____

Mixed Applications

USE DATA For 3–4, use the table.

3. Steve biked through the Appalachian Mountains for his vacation. He rode his bike for 9 miles each day until he finished his trip. How many miles did Steve bike on his last day?

____

4. If all bikers rode for 9 miles each day, who had to bike the least on the last day to finish their trip?

____

Miles Biked on Vacation

Biker	Miles
Jen	114
Steve	124
Brianna	137
Carl	109

5. Jessica has taken 7 tests this year. Her test scores are 87, 96, 91, 89, 95, 89, and 97. What is Jessica's average for these tests?

____

6. Kevin has a collection of 547 stamps. Each page of his stamp book can hold 24 stamps. How many pages will Kevin use in his stamp book?

____

Name____________________

Algebra: Patterns in Division

Use basic facts and patterns to find the quotient.

1. $60 \div 10$ ________

2. $140 \div 7$ ________

3. $180 \div 90$ ________

4. $480 \div 6$ ________

5. $400 \div 50$ ________

6. $160 \div 40$ ________

7. $360 \div 6$ ________

8. $560 \div 80$ ________

9. $240 \div 3$ ________

10. $200 \div 10$ ________

11. $630 \div 70$ ________

12. $420 \div 60$ ________

13. $810 \div 90$ ________

14. $800 \div 2$ ________

15. $900 \div 30$ ________

16. $350 \div 50$ ________

Compare. Use <, >, or = for each ◯.

17. $35 \div 7$ ◯ $350 \div 7$

18. $240 \div 80$ ◯ $24 \div 8$

19. $360 \div 4$ ◯ $360 \div 40$

20. $120 \div 3$ ◯ $120 \div 30$

21. $25 \div 5$ ◯ $250 \div 5$

22. $320 \div 80$ ◯ $32 \div 8$

Problem Solving and TAKS Prep

23. A school ordered 10 cases of paper. The paper weighed a total of 700 pounds. How much did one case of paper weigh?

____

24. An office bought 8 office chairs for a total of $720. Each chair came with a $15 mail-in rebate. After the mail-in, how much money did each chair cost?

____

25. A clothing store spends $450 on 9 clothes racks. How much does each clothes rack cost?

A $90

B $450

C $405

D $50

26. A business man spends $640 for 8 desk phones for his company. How much does each desk phone cost?

F $70

G $80

H $640

J $8

Name______________________________ **Lesson 6.2**

Estimate with 2-Digit Divisors

Write two pairs of compatible numbers for each. Then give two possible estimates.

1. $38\overline{)329}$
2. $54\overline{)386}$
3. $75\overline{)384}$
4. $63\overline{)479}$

5. $425 \div 88$
6. $543 \div 91$
7. $167 \div 26$
8. $237 \div 73$

Estimate the quotient.

9. $24\overline{)157}$
10. $31\overline{)229}$
11. $38\overline{)793}$
12. $72\overline{)681}$

13. $181 \div 35$
14. $516 \div 62$
15. $436 \div 51$
16. $593 \div 87$

Problem Solving and TAKS Prep

17. The distance from the bottom of the first floor of an office building to the top of the 86th floor is 353 meters. Estimate how many meters tall is each floor.

18. Maria ran one mile in 8 minutes after school. At the same time, Joshua ran one mile in 540 seconds. Who ran the mile in less time?

19. Joe built a tower out of blocks. It was 175 centimeters tall. The height of each cube was 8 centimeters. About how many cubes did Joe use?
    - A 10
    - B 21
    - C 18
    - D 48

20. Heather spent 480 minutes practicing basketball last month. How many hours did Heather spend practicing basketball last month?
    - F 60
    - G 4
    - H 10
    - J 8

Name____ Lesson 6.3

Divide by 2-Digit Divisors

Divide. Check your answer.

1. $23\overline{)782}$

2. $42\overline{)672}$

3. $68\overline{)816}$

4. $18\overline{)851}$

5. $36\overline{)480}$

6. $53\overline{)922}$

7. $41\overline{)869}$

8. $26\overline{)767}$

9. $897 \div 74$ ____________

10. $928 \div 65$ ____________

11. $947 \div 39$ ____________

12. $619 \div 18$ ____________

Problem Solving and TAKS Prep

13. The average person eats 53 pounds of bread each year. How many years would it take for the average person to eat 689 pounds of bread?

____

14. The average person in the U.S. uses 47 gallons of water each day. How many days would it take for the average U.S. person to use 846 gallons of water?

____

15. The school auditorium has 756 seats arranged in 27 equal rows. How many seats are in each row?

A 27

B 28

C 29

D 30

16. A farmer planted a total of 768 corn seeds in 24 equal rows. How many corn seeds are there in each row?

F 28

G 30

H 32

J 34

Name ____________________

Correcting Quotients

Write *low, high,* or *just right* for each estimate.

1. $\begin{array}{r}7\\43\overline{)527}\end{array}$ ______

2. $\begin{array}{r}12\\18\overline{)134}\end{array}$ ______

3. $\begin{array}{r}9\\34\overline{)253}\end{array}$ ______

4. $\begin{array}{r}7\\57\overline{)426}\end{array}$ ______

5. $\begin{array}{r}10\\67\overline{)915}\end{array}$ ______

Divide.

6. $18\overline{)954}$

7. $27\overline{)231}$

8. $32\overline{)681}$

9. $63\overline{)214}$

10. $79\overline{)328}$

11. $615 \div 49$ ______

12. $448 \div 56$ ______

13. $486 \div 30$ ______

14. $824 \div 62$ ______

15. $571 \div 94$ ______

Problem Solving and TAKS Prep

16. Robin needs to buy 250 coasters for a graduation party. Each package contains 18 coasters. How many packages should Robin buy?

____

17. A store ordered 832 ounces of floor cleaner. Each bottle is 32 ounces and costs $3. How much did the store spend on the order?

____

18. The Comfortable Shoe Company can fit 16 boxes of shoes in a crate. How many crates will the company need to pack 576 boxes of shoes?

A 36

B 40

C 35

D 30

19. A Disc Jockey has a collection of 816 CDs. The CD case that he likes holds 24 CDs. How many cases will the Disc Jockey need to hold all his CDs?

F 43

G 30

H 34

J 40

Name____________________ **Lesson 6.5**

Problem Solving Workshop Skill: Multistep Problems

Problem Solving Skill Practice

Solve.

1. A group of students went to the Science Museum. They spent a total of $342 on t-shirt and poster souvenirs. Each student bought 2 souvenir posters and paid for 1 t-shirt souvenir. Each poster cost $2 and each t-shirt cost $5. How many students were in the group?

____

2. Joy has a 35 page photo album which contains 833 pictures. The last page has 17 pictures. How many pictures are on each of the other pages if the pictures are equally divided among the pages?

____

Mixed Applications

USE DATA For 3–6, use the table.

3. Mr. Dorrance paid $540 to send his 4 children to summer camp. Two of his children went to Nature camp and another went to Sports camp. What camp did his fourth child attend?

____

4. **Use Data** Children who attend the Space camp are there for 6 hours each day. How much does it cost per hour?

____

Summer Camps

Camp	Cost for Five Days
Music	$140
Sports	$170
Space	$240
Nature	$115
Computer	$175

5. Mrs. Kerrigan wants to send 3 children to Nature camp and 2 children to Computer camp. How much will Mrs. Kerrigan need to pay?

____

6. Is it reasonable that the total cost for 8 children to attend Space camp to be about $2,000? Explain.

____

____

Name ______________________

Divide Money

Find the quotient.

1. $4\overline{)\$948}$ ______

2. $3\overline{)\$861}$ ______

3. $6\overline{)\$912}$ ______

4. $7\overline{)\$882}$ ______

5. $5\overline{)\$935}$ ______

6. $12\overline{)\$552}$ ______

7. $18\overline{)\$486}$ ______

8. $15\overline{)\$570}$ ______

9. $\$918 \div 2$ ______

10. $\$944 \div 8$ ______

11. $\$931 \div 19$ ______

12. $\$544 \div 32$ ______

Problem Solving and TAKS Prep

13. A theater group is selling cookies to raise money. They sold 13 cases of cookies for $975. How much does a case of cookies cost?

______

14. A book club orders 17 books, one for each member. The total cost of the book order is $102. How much money does each book cost?

______

15. Scott worked for 22 hours. He earned $286. How much did Scott earn per hour?

A $14

B $16

C $15

D $13

16. Jen worked 21 hour she earned $315. How much did Jen earn per hour?

F $19

G $13

H $15

J $12

Name________________________________ **Lesson 7.1**

Expressions and Variables

Write a numerical expression. Tell what the expression represents.

1. William shared 8 apples equally among 4 friends.

2. Jillian bought 4 toys for \$7 each.

3. 35 more than 18

Write an algebraic expression. Tell what the variable represents.

4. Jasmine has three times as many chores as her younger brother does.

5. Pedro swam some laps in the pool and then swam 2 more.

6. Neil spent 25 minutes on his math and some more time on his history homework.

Problem Solving and TAKS Prep

USE DATA For 7–8, use the table.

The rule for the number of fish for an aquarium is 1 gallon of water for each inch of fish.

Aquarium Fish

Type of Fish	Length (in inches)
Bronze Cory	3
Clown Barb	5
Silver Dollar	8

7. Write a numerical expression to represent the total number of silver dollars that could be in a 24-gallon tank. Let d = number of silver dollars.

8. Jason has 9 bronze corys in a tank. Write a numerical expression that to find the minimum number of gallons of water in the tank.

9. Kendra has some cell phone minutes. Her friend has 2 times as many cell phone minutes. Write an expression to represent the number of cell phone minutes Kendra's friend has.

10. The temperature increased from a low of 62 degrees. Which expression best describes the new temperature?

A $62 - t$

B $62 + t$

C $62t$

D $t \div 62$

Name ____________________

Evaluate Expressions

Evaluate each expression.

1. $3 \times (n + 7)$ if $n = 4$ ______

2. $(17 + 6) \times 2$ ______

3. $\frac{21}{n}$ if $n = 7$ ______

4. $13 \times n$ if $n = 4$ ______

5. $(28 \div 4) - n$ if $n = 2$ ______

6. $(12.4 - 3) - 4$ ______

7. $8n$ if $n = 4$ ______

8. $16 + (28 - 19)$ ______

9. $24.7 + (n - 6)$ if $n = 9$ ______

10. $8n - 5$ if $n = 6$ ______

11. $24 + (18 - 11)$ ______

12. $\frac{35}{n}$ if $n = 5$ ______

13. $22 - 2n$ if $n = 8$ ______

14. $(4 + n) \times 6$ if $n = 5$ ______

15. $7n + 8$ if $n = 5$ ______

16. $21 + (1.4 + 3)$ ______

Use the expression to complete each table.

17.

h	6	12	24	42
$h \div 6$				

18.

w	9	15	27	45
$w \times 3$				

Problem Solving and TAKS Prep

19. Ling paid $12 each for 5 T-shirts. She also paid $15 for a sweatshirt. Write and evaluate an expression that shows the total amount Ling paid.

____

20. Harry runs a canoe shop. He has 15 canoes he rents out. In the morning, 5 canoes are rented. In the afternoon 6 more canoes are rented. Write and evaluate an expression that shows how many canoes are still available to be rented.

____

21. Greta bought 5 bunches of flowers. There were 9 flowers in each bunch. The flowers were yellow or red. Greta counted 22 red flowers. How many flowers were yellow?

A $9 \times 5 + 22$ **C** $22 - 9 + 5$

B $9 + 5 + 22$ **D** $9 \times 5 - 22$

22. Juan rented a canoe for 3 hours at $8 per hour. He also paid $5 to rent a life vest. How much did Juan pay altogether?

F $3 \times \$8 + \$5$ **H** $3 \times \$5 \times \$8$

G $3 \times \$5 + \$8$ **J** $3 \times \$8 - \$5$

Name____

Write Equations

Write an equation for each. Tell what the variable represents.

1. Paulina has a photo album with 60 photos. Each page contains 5 photos. How many pages does tthe album have?

____

____

2. Jarrod practiced the trumpet and piano for 45 minutes. He practiced piano for 15 minutes. How long did he practice the trumpet?

____

____

Write a problem for each equation. Tell what the variable represents.

3. $7t = 63$

____

4. $6 + b = 11$

____

Problem Solving and TAKS Prep

5. Jaime has $130 in her savings account. She wants to buy a bike for $225. How much more money does Jaime need to buy the bike? Write an equation with a variable to represent the problem.

____

6. **What if** Jamie already bought the bike and has $29 left in her account. How much money did she have before buying the bike? Write an equation with a variable to represent the problem.

____

7. The Amsco building is 135 feet tall. The Tyler building is 30 feet shorter than the Amsco building. What is the Tyler building's height? Write an equation to represent this problem.

A $135 = 30h$

B $h = 135 - 30$

C $135 = 30 - h$

D $h = 135 + 30$

8. Tam had downloaded 25 songs for her MP3 player. She then downloaded some more songs. She now has 31 songs for her MP3 player. How many songs did Tam download? Write an equation to represent this problem.

F $25 + s = 31$

G $s - 31 = 25$

H $s - 25 = 31$

J $56 - s = 31$

Name____ Lesson 7.4

Solve Equations

Which of the numbers 5, 7, or 12 is the solution of the equation?

1. $t - 2 = 5$ ________
2. $30 \div e = 6$ ________
3. $3 \times u = 36$ ________
4. $18 + p = 30$ ________

Use mental math to solve each equation. Check your solution.

5. $56 = 8 \times t$ ________
6. $22 = p + 9$ ________
7. $25 - n = 13$ ________
8. $72 \div y = 12$ ________

Solve each equation. Check your solution.

9. $d \div 4 = 8$ ________
10. $6 \times s = 84$ ________
11. $v - 14 = 38$ ________
12. $\$24 + r = \$61$ ________

Find the value of each variable.

13. $h \times 3 = 60 \div 4$ ________
14. $36 - 17 = n + 5$ ________
15. $u + 7 = 64 \div 8$ ________
16. $21 + 24 = 5 \times g$ ________

Problem Solving and TAKS Prep

17. **Algebra** A bear weighed 165 pounds when it came out of hibernation. During the summer it gained *n* pounds. At the end of the summer the bear weighed 240 pounds. Write and solve an equation to find out how much the bear gained during the summer.

____

18. **Algebra** Sam took 42 pictures of animals on a nature hike. He placed the same number of pictures on each page of an album. He used 7 pages of his album. Write and solve an equation to find out how many pictures he placed on each page of his album.

____

19. The equation $\$56 \div p = \$8$ represents the total cost of some books and the cost per book. How many books were bought?

A 7

B 8

C 9

D 12

20. Jesse had a book of 14 crossword puzzles. After solving some of the puzzles, he has 3 puzzles left. Write and solve an equation to find out how many crossword puzzles Jesse solved.

____

____

Name______________________________________

Problem Solving Workshop Strategy: Draw a Diagram

Problem Solving Strategy Practice

Draw a diagram to solve.

1. Greg, Mike, and Sam each caught a fish in a fishing tournament. Greg's fish was 5 pounds heavier than Mike's. Mike's fish was 11 pounds lighter than Sam's. Greg's fish weighed 11 pounds. How much did Sam's fish weigh?

2. Cora went fishing and caught a redfish, a spotted sea trout, and a snook. The sea trout weighed twice as many pounds as the redfish. The redfish weighed twice as many pounds as the snook. Cora caught a total of 21 pounds of fish. How much did the snook weigh?

Mixed Strategy Practice

USE DATA For 3–5, use information from the table.

3. Bill caught a fish that weighs half as much as Jill's fish. Terry caught a fish that weighs three times as much as Bill's fish. How much did Terry's fish weigh?

Blue Lake Fishing Competition Results

Competitor	Weight of Fish	Length of Fish
John	9 pounds	7 inches
Matthew	18 pounds	14 inches
Rosa	15 pounds	15 inches
Jill	12 pounds	23 inches

4. The winning competitor caught a fish which weighed more than 10 pounds and was less than 15 inches long. Which competitor won the competition?

5. Two of the competitors were siblings. The combined length of their fish was 22 inches. Which two competitors were siblings?

Name ____________________

Lesson 7.6

Mental Math: Use the Properties

Use properties and mental math to find the value.

1. $12 + 18 + 39$ ______

2. $53 + 64 + 37$ ______

3. 6×103 ______

4. $(20 \times 4) \times 3$ ______

5. $41 + 29 + 46$ ______

6. $26 + 43 + 34$ ______

7. $6 \times 15 \times 2$ ______

8. 4×180 ______

9. $72 + 18 + 32$ ______

10. $7 \times 4 \times 15$ ______

11. 34×6 ______

12. $33 + (37 + 32)$ ______

13. 42×7 ______

14. $29 + 46 + 51$ ______

15. $5 \times 6 \times 12$ ______

16. 62×4 ______

17. $36 + 18 + 24$ ______

18. $12 \times 6 \times 4$ ______

Problem Solving and TAKS Prep

19. **FAST FACT** A group of sea lions together in the water are called a raft. In a raft, sea lions can safely rest together. During one afternoon, a research team saw 4 rafts of sea lions. Each raft had 16 sea lions in it. How many sea lions did the research team see?

______

20. Tell which property you would use to mentally find the value of $5 \times 4 \times 45$. Then find the value.

______

______

21. There are 6 shelving units containing 5 shelves each. Each shelf holds 35 DVDs. Find the total number of DVDs on the shelving unit.

A 210

B 450

C 950

D 1,050

22. Tickets for the movies cost $13 each. James' family buys 6 tickets. Explain how to use mental math to find the total cost of the movie tickets.

______

______

Name______________________________

Factors and Multiples

Write the common factors of each pair of numbers.

1. 3, 24 ________

2. 16, 20 ________

3. 13, 26 ________

4. 5, 10 ________

5. 22, 24 ________

Write the first three common multiples of each pair of numbers.

6. 2, 4 ________

7. 5, 3 ________

8. 8, 6 ________

9. 18, 3 ________

10. 3, 2 ________

11. 6, 12 ________

12. 4, 8 ________

13. 3, 4 ________

14. 5, 6 ________

15. 4, 7 ________

Problem Solving and TAKS Prep

USE DATA For 16-17, use the table.

Pack Of Marbles

Color Of Marble	Number Per Pack
Yellow	2
Green	4
Blue	3
Orange	6

16. What are the least numbers of packs of yellow marbles and blue marbles you have to buy to have the same number of each color of marble?

____

____

17. What are the least numbers of packs of green marbles, blue marbles, and orange marbles you have to buy to have the same number of each color of marble?

____

____

18. What is the smallest common multiple of 9 and 10 that is also a multiple of 15?

____

19. Which group shows all the numbers that are common factors of 12 and 32?

A 2, 4

B 1, 2, 4

C 1, 2, 3, 4, 6, 12

D 1, 2, 4, 8, 16

Name________________________________ **Lesson 8.2**

Greatest Common Factor

Write the GCF of each set of numbers.

1. 12, 36 ________ **2.** 21, 56 ________ **3.** 14, 21 ________ **4.** 8, 24 ________ **5.** 15, 25 ________

6. 12, 18 ________ **7.** 4, 12 ________ **8.** 4, 8, 16 ________ **9.** 6, 50, 60 ________ **10.** 9, 21, 54 ________

Problem Solving and TAKS Prep

USE DATA For 11–12, use the table.

11. Sharon is dividing her polished rock collection into bags. Each bag will contain the same number of each color of rocks. She is placing the green and blue rocks into the same bags. How many rocks will be in each bag?

____

Sharon's Rock Collection	
Polished Rocks	**Number Of Polished Rocks**
Red	12
Yellow	28
Green	16
Blue	24

12. Sharon is placing the red rocks and the yellow rocks together in the bags. Each bag will contain the same number of each color of rocks. How many bags will contain red and yellow rocks?

____

13. This list contains factors of which number?

2, 5, 25, 10, 50, 1

A 25

B 50

C 75

D 5

14. Which number is the greatest common factor of 5 and 25?

F 1

G 5

H 15

J 25

Name________________________________ Lesson 8.3

Prime and Composite Numbers

Write *prime* or *composite*. You may use counters or draw arrays.

1. 12 ______ **2.** 37 ______ **3.** 44 ______

4. 28 ______ **5.** 35 ______ **6.** 122 ______

7. 61 ______ **8.** 72 ______ **9.** 89 ______

10. 56 ______ **11.** 49 ______ **12.** 59 ______

13. 101 ______ **14.** 75 ______ **15.** 88 ______

16. 14 ______ **17.** 83 ______ **18.** 109 ______

Name______________________________ Lesson 8.4

Problem Solving Workshop Strategy: Make an Organized List

Problem Solving Strategy Practice

Use an organized list to solve.

1. During the month of May, Jean has photography class every third day and a photography show every Saturday. On May 5 she has class and a show. During the month of May, how many more times will she have a class and a show on the same day?

____

2. Students are making picture frames. They can choose from a brown or black picture frame and a red, yellow, blue, or green mat. How many different picture frame/mat combinations can the students make?

____

Mixed Strategy Practice

USE DATA For 3–5, use the graph.

3. Robin has 7 red beads, 27 purple beads, and 24 yellow beads. She wants to make a necklace with the pattern: 1 red bead; 3 purple beads; 2 yellow beads. How many times can she repeat the pattern? Which color of beads will she run out of first?

____

4. Clay asked his classmates about their favorite type of book. After making the graph, he accidentally spilled water on it. What information is missing from the graph?

____

____

____

5. Complete the graph. Use the clues below to find the missing data in the graph.

Clue 1: The least favorite type of book is fantasy.

Clue 2: Mystery books are favored by 10% more students than western books.

Which Type Of Book Is Your Favorite

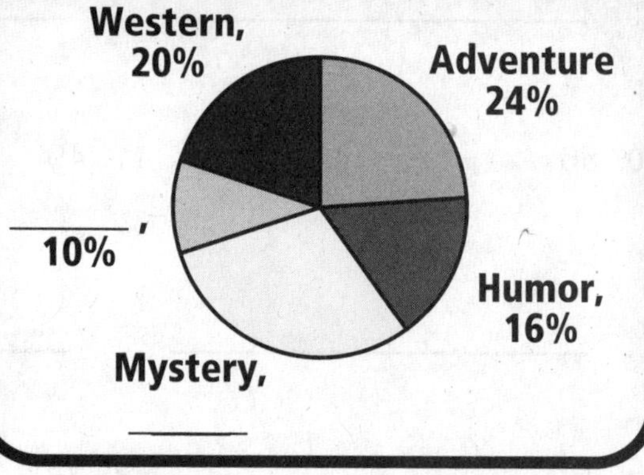

Name______________________________

Number Relationships

Prime Numbers from 2 to 223											
2	3	5	7	11	13	17	19	23	29	31	37
41	43	47	53	59	61	67	71	73	79	83	89
97	101	103	107	109	113	127	131	137	139	149	151
157	163	167	173	179	181	191	193	197	199	211	223

For 1–28, use the table. Tell whether adding 2 to each prime number will result in another prime number.

1. 2 ____________

2. 7 ____________

3. 73 ____________

4. 13 ____________

5. 43 ____________

6. 3 ____________

7. 11 ____________

8. 67 ____________

9. 83 ____________

10. 79 ____________

11. 59 ____________

12. 19 ____________

13. 89 ____________

14. 181 ____________

15. 47 ____________

16. 149 ____________

17. 113 ____________

18. 61 ____________

19. 97 ____________

20. 103 ____________

21. 109 ____________

22. 197 ____________

23. 71 ____________

24. 101 ____________

25. 173 ____________

26. 127 ____________

27. 211 ____________

28. 191 ____________

Name________________________________ Lesson 9.1

Equivalent Fractions

Write an equivalent fraction.

1. $\frac{1}{8}$ ______ 2. $\frac{7}{10}$ ______ 3. $\frac{4}{5}$ ______ 4. $\frac{6}{8}$ ______ 5. $\frac{3}{4}$ ______ 6. $\frac{1}{3}$ ______

7. $\frac{3}{6}$ ______ 8. $\frac{8}{12}$ ______ 9. $\frac{6}{9}$ ______ 10. $\frac{10}{15}$ ______ 11. $\frac{10}{16}$ ______ 12. $\frac{5}{6}$ ______

13. $\frac{2}{4}$ ______ 14. $\frac{3}{12}$ ______ 15. $\frac{4}{6}$ ______ 16. $\frac{4}{10}$ ______ 17. $\frac{1}{5}$ ______ 18. $\frac{12}{16}$ ______

Problem Solving and TAKS Prep

USE DATA For 19–20, use the table.

19. Natalie asked people which of the six colors in the chart they liked more than the rest. Write four equivalent fractions to show the fraction of people who chose red.

____

20. Natalie asks 4 more people their opinion,and they all say blue. What three equivalent fractions show the fraction of people who chose red?

____

Preferred Colors

Color	Number of People Who Chose It
Orange	1
Red	4
Purple	2
Blue	3
Green	1
Yellow	1

21. Which fraction is equivalent to $\frac{2}{5}$?

A $\frac{3}{10}$ C $\frac{7}{10}$

B $\frac{4}{10}$ D $\frac{3}{5}$

22. Which fraction is equivalent to $\frac{14}{16}$?

F $\frac{7}{8}$ H $\frac{4}{6}$

G $\frac{7}{9}$ J $\frac{2}{16}$

Name____________________ Lesson 9.2

Simplest Form

Name the GCF of the numerator and denominator.

1. $\frac{14}{16}$ ________ **2.** $\frac{4}{4}$ ________ **3.** $\frac{12}{35}$ ________ **4.** $\frac{9}{30}$ ________ **5.** $\frac{10}{25}$ ________

6. $\frac{8}{22}$ ________ **7.** $\frac{17}{34}$ ________ **8.** $\frac{28}{77}$ ________ **9.** $\frac{16}{100}$ ________ **10.** $\frac{24}{30}$ ________

Write each fraction in simplest form.

11. $\frac{10}{10}$ ________ **12.** $\frac{9}{16}$ ________ **13.** $\frac{20}{60}$ ________ **14.** $\frac{36}{45}$ ________ **15.** $\frac{12}{57}$ ________

16. $\frac{10}{24}$ ________ **17.** $\frac{15}{25}$ ________ **18.** $\frac{32}{40}$ ________ **19.** $\frac{70}{100}$ ________ **20.** $\frac{48}{60}$ ________

Problem Solving and TAKS Prep

21. **Fast Fact** Eight states border one or more of the five Great Lakes. Write a fraction representing the part of the 50 states that border a Great Lake. Write the fraction in simplest form.

____

22. Twenty out of 75 salon clients made an appointment for a haircut. What fraction of the clients made a haircut appointment? Write the fraction in simplest form.

____

23. Which fraction shows $\frac{21}{28}$ in simplest form?

A $\frac{1}{8}$

B $\frac{1}{7}$

C $\frac{3}{7}$

D $\frac{3}{4}$

24. Twelve of 30 students rode the bus today. What fraction of the students rode the bus? Write the fraction in simplest form.

____

Name____ **Lesson 9.3**

Model Mixed Numbers

Write a mixed number and a fraction for each.

1.

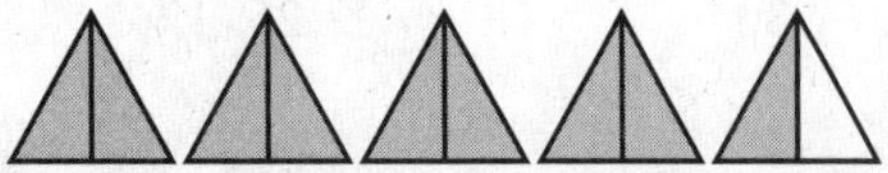

____

2.

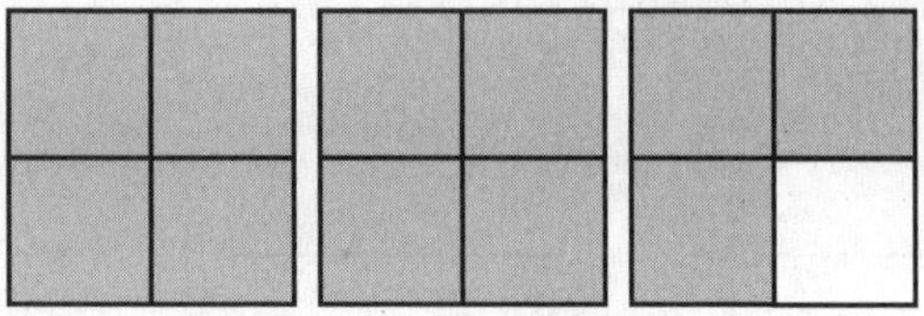

____

3.

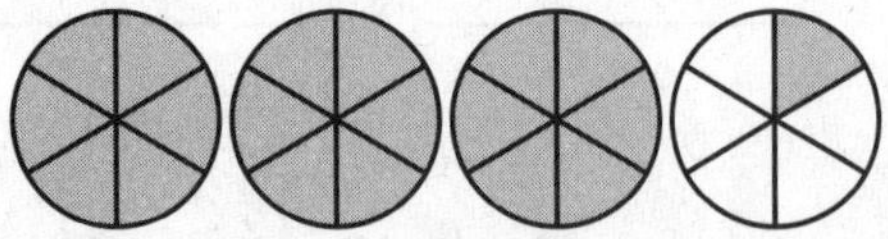

____

4.

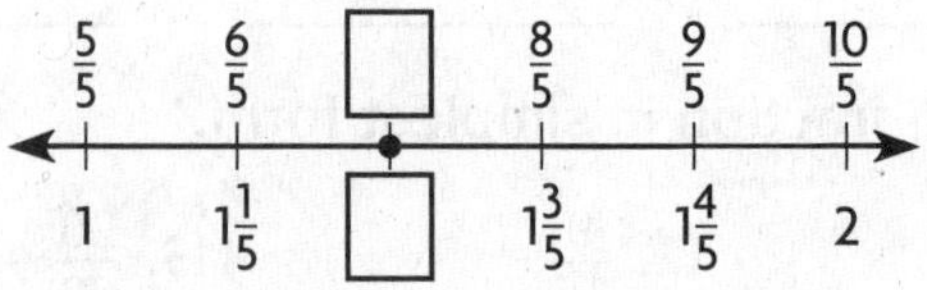

____

5.

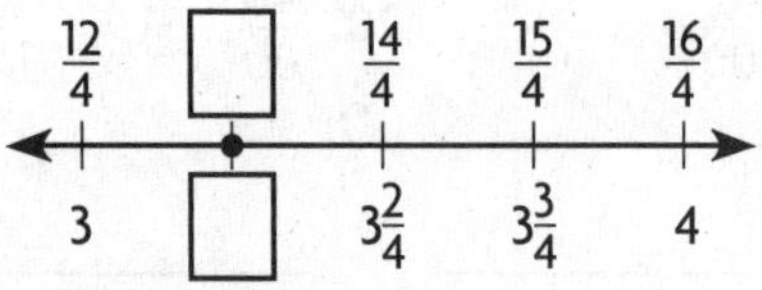

____

6.

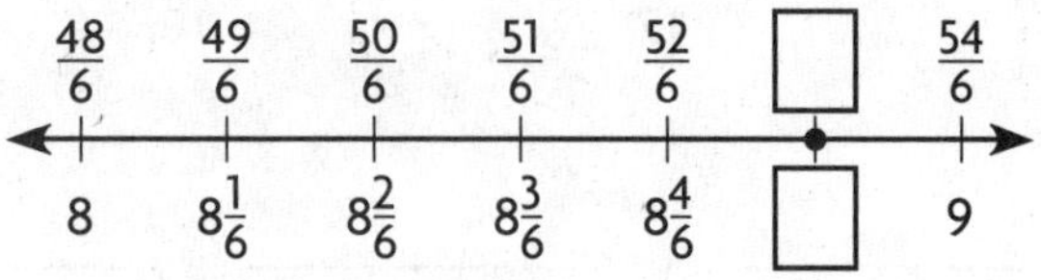

____

Use the number line to identify each pair of numbers.
Tell whether each pair is *equivalent* or *not equivalent*.

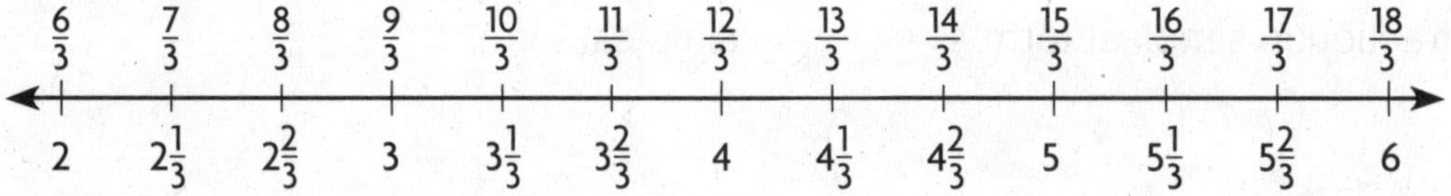

7. 6 and $\frac{18}{3}$ ______________

8. $5\frac{2}{3}$ and $\frac{17}{3}$ ______________

9. $2\frac{2}{3}$ and $\frac{7}{3}$ ______________

10. $3\frac{1}{3}$ and $\frac{10}{3}$ ______________

11. $4\frac{2}{3}$ and $\frac{14}{3}$ ______________

12. $5\frac{1}{3}$ and $\frac{16}{3}$ ______________

13. 2 and $\frac{12}{3}$ ______________

14. $3\frac{2}{3}$ and $\frac{11}{3}$ ______________

Name ______________________________ Lesson 9.4

Rename Fractions and Mixed Numbers

Write each mixed number as a fraction. Write each fraction as a mixed number.

1. $1\frac{7}{8}$ ________ **2.** $\frac{10}{9}$ ________ **3.** $\frac{13}{4}$ ________ **4.** $3\frac{4}{5}$ ________ **5.** $1\frac{1}{10}$ ________ **6.** $4\frac{1}{6}$ ________

7. $\frac{21}{10}$ ________ **8.** $\frac{19}{8}$ ________ **9.** $\frac{11}{3}$ ________ **10.** $2\frac{9}{10}$ ________ **11.** $3\frac{1}{9}$ ________ **12.** $\frac{18}{5}$ ________

13. $1\frac{3}{7}$ ________ **14.** $\frac{9}{4}$ ________ **15.** $\frac{17}{7}$ ________ **16.** $1\frac{5}{6}$ ________ **17.** $2\frac{2}{9}$ ________ **18.** $\frac{7}{3}$ ________

Problem Solving and TAKS Prep

19. How many times will Gayle fill a $\frac{1}{2}$-cup ladel to serve $8\frac{1}{2}$ cups of punch?

____

20. A recipe calls for $2\frac{3}{4}$ cups of milk. Write $2\frac{3}{4}$ as a fraction.

____

21. Daryl uses $2\frac{1}{4}$ cups of flour to make muffins. Which fraction is equivalent to $2\frac{1}{4}$?

A $\frac{5}{4}$

B $\frac{6}{4}$

C $\frac{9}{4}$

D $\frac{21}{4}$

22. Lana needs $3\frac{2}{3}$ yards of ribbon for a costume. Which fraction is equivalent to $3\frac{2}{3}$?

F $\frac{5}{3}$

G $\frac{7}{3}$

H $\frac{11}{3}$

J $\frac{32}{3}$

Name____ Lesson 9.5

Compare and Order Fractions and Mixed Numbers

Compare. Write <, >, or = for each ○.

1. $\frac{4}{9}$ ○ $\frac{5}{9}$
2. $\frac{3}{4}$ ○ $\frac{3}{5}$
3. $\frac{2}{3}$ ○ $\frac{8}{12}$
4. $\frac{5}{8}$ ○ $\frac{4}{7}$
5. $\frac{9}{11}$ ○ $\frac{8}{9}$
6. $\frac{5}{12}$ ○ $\frac{3}{7}$
7. $\frac{6}{10}$ ○ $\frac{4}{5}$
8. $2\frac{7}{9}$ ○ $2\frac{5}{6}$
9. $4\frac{5}{8}$ ○ $4\frac{3}{4}$
10. $9\frac{2}{6}$ ○ $8\frac{3}{9}$
11. $3\frac{4}{5}$ ○ $3\frac{5}{6}$
12. $1\frac{2}{10}$ ○ $1\frac{1}{5}$
13. $4\frac{4}{6}$ ○ $3\frac{3}{4}$
14. $1\frac{1}{3}$ ○ $1\frac{4}{12}$
15. $6\frac{3}{8}$ ○ $6\frac{1}{4}$
16. $7\frac{5}{6}$ ○ $9\frac{5}{6}$
17. $2\frac{4}{9}$ ○ $2\frac{1}{5}$
18. $5\frac{3}{4}$ ○ $5\frac{2}{3}$
19. $7\frac{4}{6}$ ○ $8\frac{1}{2}$
20. $1\frac{5}{11}$ ○ $1\frac{3}{7}$

Write in order from least to greatest.

21. $\frac{3}{8}, \frac{3}{4}, \frac{1}{4}$ ______
22. $\frac{2}{3}, \frac{1}{6}, \frac{7}{9}$ ______
23. $1\frac{5}{8}, 1\frac{3}{4}, 1\frac{5}{6}$ ______
24. $7\frac{3}{5}, 6\frac{2}{3}, 6\frac{6}{10}$ ______

Problem Solving and TAKS Prep

USE DATA For 25–26, use the table.

Len's Flutes	
Flute Name	**Length, in inches**
Lily	$6\frac{3}{4}$
Rose	$6\frac{5}{8}$
Ivy	$6\frac{7}{12}$

25. Len paints and sells wooden flutes. List the flutes in order from shortest to longest.

______

26. Len created a new flute that is $6\frac{2}{3}$ inches long. Which, if any, of his flutes are longer?

______

27. Kayla practiced violin $2\frac{1}{4}$ hours on Monday, $1\frac{3}{10}$ hours on Tuesday, and $1\frac{4}{9}$ hours on Wednesday. On which day did she practice the shortest?

______

28. Dean practiced trombone $1\frac{2}{3}$ hours on Monday, $1\frac{7}{12}$ hours on Tuesday, and $1\frac{7}{9}$ hours on Wednesday. On which day did he practice the longest?

______

Name________________________________ **Lesson 9.6**

Problem Solving Strategy: Make a Model

Problem Solving Strategy Practice

Make a model to solve the problem.

1. From home, Todd walked 3 blocks south and 2 blocks east to a friend's house. Then they walked 6 blocks west to school. He cannot cut across blocks. How many blocks from school does Todd live?

____

2. Kalere is erecting a picket fence on one side of her garden. Each picket is 4 inches wide and 2 inches apart. She has 12 pickets. How many inches long will Kalere's fence be?

____

Mixed Applications

Solve.

3. Lisa spent 10 minutes driving to the grocery store and 50 minutes shopping there. She spent 10 minutes driving back home and 40 minutes making sandwiches for a picnic. She drove 30 minutes from home and arrived at the picnic at 3:30 P.M. What time did Lisa leave to go to the grocery store?

____

4. Playing golf, Leta's ball stopped $3\frac{5}{8}$ feet from the hole, Blake's ball stopped $3\frac{2}{3}$ feet from the hole, and Toby's ball stopped $4\frac{1}{4}$ feet from the hole. Whose ball was closest to the hole?

____

5. A city garden is in the shape of a rectangle. There is a walkway from each corner of the rectangle to every other corner of the rectangle. How many walkways are there?

____

6. **Pose a Problem** Look back at Exercise 5. Write a similar problem by increasing the number of corners the garden has. Then solve the problem.

____

____

Name ____________________ **Lesson 9.7**

Relate Fractions and Decimals

Write each fraction as a decimal.

1. $\frac{7}{100}$ **2.** $\frac{1}{4}$ **3.** $\frac{3}{10}$ **4.** $\frac{9}{20}$ **5.** $\frac{20}{25}$ **6.** $\frac{6}{25}$

7. $\frac{2}{5}$ **8.** $\frac{1}{20}$ **9.** $\frac{13}{50}$ **10.** $\frac{10}{20}$ **11.** $\frac{14}{50}$ **12.** $\frac{2}{4}$

Write each decimal as a fraction.

13. 0.59 **14.** 0.06 **15.** 0.7 **16.** 0.41 **17.** 0.90

18. 0.05 **19.** 0.5 **20.** 0.23 **21.** 0.75 **22.** 0.08

23. 0.2 **24.** 0.22 **25.** 0.04 **26.** 0.98 **27.** 0.25

Problem Solving and TAKS Prep

28. Write a decimal for the shaded part.

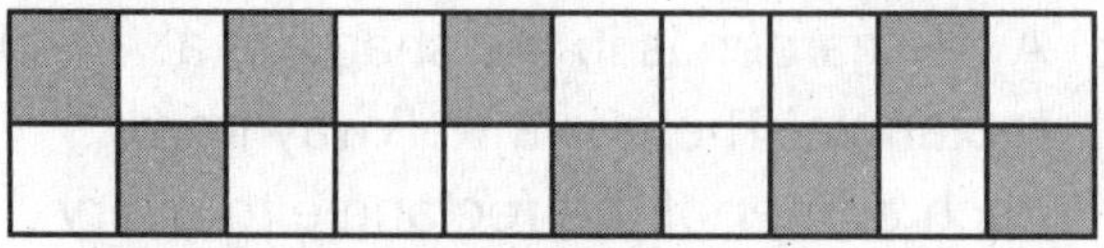

29. Which decimal is equivalent to $\frac{3}{20}$?

A 3.20

B 2.3

C 0.3

D 0.15

30. Which is equivalent to 1.8?

F $1\frac{4}{5}$

G $1\frac{1}{8}$

H $\frac{1}{8}$

J $\frac{1}{18}$

Name___________________________________

Model Fraction Addition

Use fraction bars to add. Write the sum in simplest form.

1. $\frac{1}{4} + \frac{1}{4}$ ________
2. $\frac{3}{8} + \frac{1}{8}$ ________
3. $\frac{2}{5} + \frac{1}{5}$ ________
4. $\frac{3}{7} + \frac{2}{7}$ ________
5. $\frac{7}{10} + \frac{2}{10}$ ________
6. $\frac{3}{8} + \frac{3}{8}$ ________
7. $\frac{1}{11} + \frac{5}{11}$ ________
8. $\frac{1}{6} + \frac{2}{6}$ ________
9. $\frac{4}{9} + \frac{2}{9}$ ________
10. $\frac{1}{5} + \frac{3}{5}$ ________
11. $\frac{3}{10} + \frac{5}{10}$ ________
12. $\frac{9}{12} + \frac{2}{12}$ ________
13. $\frac{4}{7} + \frac{2}{7}$ ________
14. $\frac{1}{5} + \frac{3}{5}$ ________
15. $\frac{5}{12} + \frac{6}{12}$ ________

Draw a picture to add. Write the sum in simplest form.

16. $\frac{1}{4} + \frac{2}{4}$ ________
17. $\frac{1}{8} + \frac{4}{8}$ ________
18. $\frac{5}{10} + \frac{2}{10}$ ________
19. $\frac{1}{5} + \frac{1}{5}$ ________
20. $\frac{7}{12} + \frac{3}{12}$ ________
21. $\frac{3}{6} + \frac{2}{6}$ ________
22. $\frac{2}{7} + \frac{4}{7}$ ________
23. $\frac{5}{8} + \frac{2}{8}$ ________
24. $\frac{7}{11} + \frac{2}{11}$ ________
25. $\frac{3}{9} + \frac{2}{9}$ ________
26. $\frac{5}{12} + \frac{4}{12}$ ________
27. $\frac{1}{7} + \frac{5}{7}$ ________
28. $\frac{3}{10} + \frac{4}{10}$ ________
29. $\frac{4}{6} + \frac{1}{6}$ ________
30. $\frac{3}{5} + \frac{1}{5}$ ________
31. $\frac{4}{9} + \frac{3}{9}$ ________
32. $\frac{1}{3} + \frac{1}{3}$ ________
33. $\frac{8}{11} + \frac{2}{11}$ ________
34. $\frac{6}{12} + \frac{1}{12}$ ________
35. $\frac{5}{9} + \frac{3}{9}$ ________

Name________________________________

Record Addition of Fractions

Add. Write the sum in simplest form.

1. $\frac{1}{8} + \frac{3}{8}$

2. $\frac{2}{9} + \frac{4}{9}$

3. $\frac{1}{6} + \frac{1}{6}$

4. $\frac{1}{10} + \frac{7}{10}$

5. $\frac{4}{11} + \frac{6}{11}$

6. $\frac{1}{8} + \frac{1}{8}$

7. $\frac{1}{4} + \frac{1}{4}$ ________

8. $\frac{2}{11} + \frac{3}{11}$ ________

9. $\frac{7}{9} + \frac{1}{9}$ ________

10. $\frac{4}{10} + \frac{3}{10}$ ________

11. $\frac{1}{5} + \frac{3}{5}$ ________

12. $\frac{2}{7} + \frac{3}{7}$ ________

Problem Solving and TAKS Prep

13. Mark uses $\frac{2}{5}$ cup of flour and $\frac{1}{5}$ cup of water in his muffin recipe. How much flour and water does Mark use in all?

____

14. **What if** Mark adds another $\frac{1}{5}$ cup of water to the muffin recipe? How many cups of water would Mark use in all?

____

15. Rodrigo used $\frac{3}{5}$ of his allowance to buy a model airplane and $\frac{1}{5}$ of his allowance to buy a comic book. How much of his allowance did Rodrigo use to buy both items?

A $\frac{2}{5}$

B $\frac{4}{5}$

C $\frac{3}{10}$

D $\frac{1}{10}$

16. Jenny studied math for $\frac{1}{3}$ hour and her spelling words for $\frac{1}{3}$ hour. How long did Jenny study her math and spelling words?

F $\frac{1}{6}$ hour

G $\frac{5}{6}$ hour

H $\frac{2}{9}$ hour

J $\frac{2}{3}$ hour

Name____________________

Model Mixed Number Addition

Use fraction bars to add. Write the sum in simplest form.

1. $1\frac{2}{5} + 1\frac{1}{5}$ ________

2. $2\frac{1}{10} + 1\frac{7}{10}$ ________

3. $1\frac{4}{7} + 1\frac{2}{7}$ ________

4. $3\frac{2}{12} + 2\frac{8}{12}$ ________

5. $1\frac{2}{4} + 2\frac{1}{4}$ ________

6. $1\frac{1}{3} + 1\frac{2}{3}$ ________

7. $2\frac{5}{9} + 2\frac{3}{9}$ ________

8. $2\frac{3}{6} + 3\frac{1}{6}$ ________

9. $1\frac{7}{11} + 2\frac{2}{11}$ ________

10. $3\frac{5}{7} + 1\frac{1}{7}$ ________

11. $1\frac{3}{8} + 2\frac{7}{8}$ ________

12. $3\frac{2}{5} + 2\frac{2}{5}$ ________

Draw a picture to add. Write the sum in simplest form.

13. $\frac{2}{7} + 1\frac{3}{7}$ ________

14. $1\frac{3}{8} + 1\frac{5}{8}$ ________

15. $1\frac{3}{9} + 2\frac{5}{9}$ ________

16. $2\frac{8}{11} + 1\frac{2}{11}$ ________

17. $2\frac{1}{5} + \frac{3}{5}$ ________

18. $3\frac{4}{10} + 3\frac{3}{10}$ ________

19. $2\frac{5}{7} + 3\frac{4}{7}$ ________

20. $1\frac{3}{6} + \frac{1}{6}$ ________

21. $2\frac{1}{4} + 2\frac{2}{4}$ ________

22. $\frac{7}{12} + 3\frac{2}{12}$ ________

23. $1\frac{1}{9} + 2\frac{6}{9}$ ________

24. $3\frac{6}{7} + 4\frac{2}{7}$ ________

25. $2\frac{7}{8} + 3\frac{5}{8}$ ________

26. $3\frac{1}{3} + 1\frac{1}{3}$ ________

27. $3\frac{3}{4} + 2\frac{2}{4}$ ________

28. $1\frac{1}{8} + 1\frac{4}{8}$ ________

Name____________________ Lesson 10.4

Record Addition of Mixed Numbers

Find and record the sum. Write it in simplest form.

1. $3\frac{5}{8} + 4\frac{4}{8}$

2. $1\frac{4}{5} + 2\frac{3}{5}$

3. $5\frac{7}{9} + 2\frac{4}{9}$

4. $4\frac{3}{12} + 2\frac{7}{12}$

5. $3\frac{3}{4} + 2\frac{3}{4}$

6. $2\frac{5}{6} + 5\frac{1}{6}$

7. $3\frac{5}{8} + 5\frac{7}{8}$

8. $1\frac{9}{11} + 7\frac{1}{11}$

9. $2\frac{4}{7} + 2\frac{6}{7}$

10. $4\frac{2}{3} + 2\frac{2}{3}$

11. $3\frac{2}{10} + 4\frac{7}{10}$

12. $1\frac{6}{11} + 3\frac{7}{11}$

13. $4\frac{2}{5} + 4\frac{4}{5}$

14. $2\frac{5}{9} + 1\frac{8}{9}$

15. $5\frac{1}{7} + 8\frac{5}{7}$

Problem Solving and TAKS Prep

16. Over a long weekend, Marissa reads magazines for $3\frac{2}{5}$ hours and reads a book for $4\frac{1}{5}$ hours. How many hours does Marissa spend reading over the long weekend?

17. Jeremy ran $2\frac{1}{4}$ miles on Friday, $3\frac{3}{4}$ on Saturday, and 4 miles on Sunday. How many miles did Jeremy run in all three days?

18. Samantha has one suitcase that weighs $4\frac{2}{5}$ pounds and another that weighs $2\frac{1}{5}$ pounds. What was their total weight?

    A $6\frac{3}{10}$ pounds **C** $2\frac{1}{5}$ pounds

    B $6\frac{3}{5}$ pounds **D** $6\frac{1}{10}$ pounds

19. Before his run, Jeremy drank $1\frac{4}{7}$ cups of water. After his run, he drank $2\frac{6}{7}$ cups of water. How much water did he drink in all?

    F $3\frac{4}{7}$ cups **H** $4\frac{3}{7}$ cups

    G $3\frac{5}{7}$ cups **J** $4\frac{5}{7}$ cups

Name_______________________________________ Lesson 10.5

Problem Solving Workshop Strategy: Guess and Check

Problem Solving Strategy Practice

Use *guess and check* to solve.

1. Rose uses mushrooms and onions in her spaghetti sauce recipe. She uses $1\frac{1}{2}$ more cups of mushrooms than onions. If she fills a $5\frac{1}{2}$ cup container with mushrooms and onions, how many cups of each ingredient did she use?

____

2. Kelli is a chef and she enjoys reading cookbooks. She read $10\frac{1}{3}$ pages on Saturday and Sunday. Kelli read $2\frac{1}{3}$ more pages on Saturday than she did on Sunday. How many pages did she read each day?

____

Mixed Strategy Practice

USE DATA For 3–4, use the table.

Fruit Punch	
Cranberry Juice	$1\frac{1}{4}$ cup
Orange Juice	$1\frac{3}{4}$ cup
Ginger Ale	$2\frac{1}{4}$ cups

3. Toby changes the recipe for fruit punch so that she uses $\frac{2}{4}$ more cup of ginger ale. How much ginger ale will Toby need for the fruit punch recipe?

____

4. Shannon wants to make the fruit punch for a party. Write in simplest form how many cups of fruit punch the recipe will make.

____

5. Julio is making burritos for a family picnic. He uses $1\frac{1}{5}$ more pounds of beans than chicken. If he uses $7\frac{3}{5}$ pounds of beans and chicken, how many pounds of each ingredient did he use?

____

6. Nina is a painter. She used $5\frac{1}{2}$ pints of red and green paint for her most recent picture. Nina used $1\frac{1}{2}$ pint more red than green paint. How many pints of each color did Nina use?

____

Name ____________________

Subtract Fractions

Use fraction bars to subtract. Write the answer in simplest form.

1. $\frac{2}{5} - \frac{1}{5}$ ______
2. $\frac{5}{7} - \frac{4}{7}$ ______
3. $\frac{4}{9} - \frac{3}{9}$ ______
4. $\frac{4}{6} - \frac{1}{6}$ ______
5. $\frac{8}{10} - \frac{5}{10}$ ______
6. $\frac{9}{12} - \frac{3}{12}$ ______
7. $\frac{2}{4} - \frac{1}{4}$ ______
8. $\frac{7}{8} - \frac{5}{8}$ ______
9. $\frac{10}{11} - \frac{3}{11}$ ______
10. $\frac{4}{5} - \frac{2}{5}$ ______
11. $\frac{7}{9} - \frac{1}{9}$ ______
12. $\frac{4}{10} - \frac{3}{10}$ ______

Draw a picture to subtract. Write the answer in simplest form.

13. $\frac{8}{11} - \frac{5}{11}$ ______
14. $\frac{6}{9} - \frac{5}{9}$ ______
15. $\frac{2}{3} - \frac{1}{3}$ ______
16. $\frac{3}{5} - \frac{1}{5}$ ______
17. $\frac{11}{12} - \frac{7}{12}$ ______
18. $\frac{6}{7} - \frac{3}{7}$ ______
19. $\frac{3}{4} - \frac{1}{4}$ ______
20. $\frac{9}{10} - \frac{5}{10}$ ______
21. $\frac{3}{6} - \frac{2}{6}$ ______
22. $\frac{4}{7} - \frac{3}{7}$ ______
23. $\frac{6}{11} - \frac{2}{11}$ ______
24. $\frac{5}{12} - \frac{3}{12}$ ______
25. $\frac{5}{6} - \frac{2}{6}$ ______
26. $\frac{6}{10} - \frac{2}{10}$ ______
27. $\frac{5}{8} - \frac{3}{8}$ ______
28. $\frac{3}{9} - \frac{2}{9}$ ______

Name ____________________ Lesson 11.2

Record Subtraction of Fractions

Subtract. Write the answer in simplest form.

1. $\frac{7}{8} - \frac{5}{8}$ ______

2. $\frac{5}{9} - \frac{4}{9}$ ______

3. $\frac{4}{6} - \frac{2}{6}$ ______

4. $\frac{8}{10} - \frac{3}{10}$ ______

5. $\frac{6}{7} - \frac{1}{7}$ ______

6. $\frac{3}{4} - \frac{2}{4}$ ______

7. $\frac{7}{11} - \frac{5}{11}$ ______

8. $\frac{3}{8} - \frac{2}{8}$ ______

9. $\frac{11}{12} - \frac{9}{12}$ ______

10. $\frac{8}{9} - \frac{5}{9}$ ______

11. $\frac{9}{20} - \frac{7}{20}$ ______

12. $\frac{5}{6} - \frac{3}{6}$ ______

13. $\frac{9}{10} - \frac{2}{10}$ ______

14. $\frac{4}{5} - \frac{1}{5}$ ______

15. $\frac{4}{7} - \frac{1}{7}$ ______

Problem Solving and TAKS Prep

16. Marisol walks her puppy $\frac{4}{9}$ mile on Monday and $\frac{7}{9}$ mile on Tuesday. How much farther did Marisol walk her puppy on Tuesday than she did on Monday?

______

17. Christopher gives his puppy $\frac{3}{7}$ of a bacon treat in the morning and $\frac{6}{7}$ of a bacon treat in the evening. How much more of the bacon treat does the puppy get in the evening than in the morning?

______

18. Katherine used $\frac{3}{7}$ of her allowance to buy a dog collar and $\frac{5}{7}$ of her allowance to buy a squeaky toy. How much more of her allowance did Katherine spend on the squeaky toy?

A $\frac{1}{7}$

B $1\frac{1}{7}$

C $\frac{2}{7}$

D $2\frac{1}{7}$

19. Melanie spent $\frac{3}{10}$ of the afternoon playing with her kitty and $\frac{8}{10}$ of the afternoon doing homework. How much more time did Melanie spend doing homework?

F $\frac{1}{4}$

G $\frac{1}{2}$

H $\frac{5}{20}$

J $1\frac{1}{10}$

Name ______________________

Subtraction with Renaming

Find the difference. Write it in simplest form.

1. $4\frac{3}{7} - 2\frac{5}{7}$ ______

2. $3\frac{5}{12} - 1\frac{7}{12}$ ______

3. $7\frac{3}{8} - 4\frac{6}{8}$ ______

4. $5 - 1\frac{6}{9}$ ______

5. $6\frac{1}{6} - 4\frac{5}{6}$ ______

6. $9 - 2\frac{9}{10}$ ______

7. $8\frac{5}{8} - 1\frac{6}{8}$ ______

8. $2 - 1\frac{1}{4}$ ______

9. $6\frac{9}{20} - 3\frac{12}{20}$ ______

10. $10 - 4\frac{3}{5}$ ______

11. $11\frac{4}{9} - 7\frac{7}{9}$ ______

12. $4 - 2\frac{2}{4}$ ______

13. $6 - \frac{3}{4}$ ______

14. $12\frac{8}{15} - 9\frac{9}{15}$ ______

15. $7\frac{1}{5} - 6\frac{4}{5}$ ______

Problem Solving and TAKS Prep

16. Ronald is helping to paint scenery. He uses $2\frac{2}{3}$ gallons of red paint from a 3 gallon container. How many gallons of paint dose Ronald have left?

______

17. Maggie is making costumes for the play. She must make a cape that is $5\frac{1}{3}$ yards long. The fabric that she is using to make the cape is 9 yards long. How much fabric will be left after Maggie makes the cape?

______

18. Juan must dance $10\frac{5}{6}$ feet across the stage floor. The stage floor is 15 feet long. How many more feet does Juan have to dance to cross the floor?

A $5\frac{1}{6}$ feet
B $4\frac{1}{6}$ feet
C $5\frac{5}{6}$ feet
D $4\frac{5}{6}$ feet

19. Casey read $6\frac{8}{9}$ scenes in the play on Monday. There are 10 scenes in the play. How many more scenes does Casey have to read to finish the play?

F $4\frac{1}{9}$
G $4\frac{1}{10}$
H $3\frac{1}{10}$
J $3\frac{1}{9}$

Name____________________

Practice with Mixed Numbers

Subtract. Write the answer in simplest form.

1. $7 - 3\frac{5}{12}$

2. $6\frac{1}{6} - 1\frac{2}{6}$

3. $12\frac{5}{7} - 5\frac{3}{7}$

4. $5\frac{4}{11} - 3\frac{7}{11}$

5. $8\frac{3}{8} - 4\frac{4}{8}$

6. $9 - 3\frac{1}{6}$

7. $6\frac{1}{5} - 3\frac{4}{5}$

8. $7\frac{1}{4} - 3\frac{1}{4}$

9. $5\frac{1}{6} - 2\frac{5}{6}$

10. $9 - 2\frac{1}{4}$

11. $4\frac{3}{5} - 2\frac{1}{5}$

12. $10\frac{1}{6} - 7\frac{3}{6}$

13. $6 - 3\frac{5}{7}$

14. $4\frac{6}{10} - 2\frac{4}{10}$

15. $3\frac{2}{9} - 2\frac{8}{9}$

16. $7\frac{11}{12} - 5\frac{10}{12}$

Problem Solving and TAKS Prep

17. A geologist weighs two different types of rocks. The quartz weighs 5 pounds. The slate is $2\frac{3}{5}$ pounds. How much heavier is the quartz than the slate?

18. Johnny brings $2\frac{1}{8}$ gallons of his favorite sports drink to the game to share. The team drinks $1\frac{7}{8}$ gallons of the sports drink. How many gallons of the sports drink are left?

19. Blake went to visit her aunt and her grandmother in Boston. Her aunt's house is 20 miles from the train station. Her grandmother's house is $8\frac{7}{12}$ miles from the train station. How many more miles is her aunt's house than her grandmother's from the train station?

A $11\frac{5}{12}$

B $12\frac{7}{12}$

C $12\frac{5}{12}$

D $11\frac{5}{20}$

20. While visiting her grandmother, Blake worked in her vegetable garden. She picked a carrot that was $8\frac{3}{4}$ inches long and a cucumber that was $7\frac{2}{4}$ long. How much longer was the carrot than the cucumber?

F $2\frac{3}{4}$

G $1\frac{3}{4}$

H $1\frac{1}{4}$

J $2\frac{3}{8}$

Name________________________________ Lesson 11.5

Problem Solving Workshop Strategy: Work Backward

Problem Solving Strategy Practice

1. Rodney's class is making a float for the Independence Day parade. They used a total of 4 yards of red fabric, white fabric, and blue fabric to decorate the float. They used $1\frac{1}{6}$ yards of red fabric and $1\frac{5}{6}$ yards of blue fabric. If the remainder of the fabric was white, how many yards of white fabric did Rodney's class use?

____

2. At the Independence Day parade, Sing used his allowance to buy several souvenirs. He paid $22.00 for two T-shirts and a baseball cap. The baseball cap cost $6.00. He does not remember the exact price of the T-shirts. How much did Sing pay for each T-shirt?

____

Mixed Strategy Practice

USE DATA For 3–4, use the table.

Materials for Parade Float	
Materials	**Amount**
Wood	$36\frac{1}{4}$ feet
Streamers	$32\frac{3}{5}$ feet
Paint	$9\frac{1}{6}$ gallons

3. Students used $8\frac{1}{4}$ feet of streamers for the front of the float and $9\frac{3}{4}$ feet of streamers for the back of the float. How many feet of streamers were left for the sides of the float?

____

4. Students used wood to build 5 pillars on the float. Each pillar used $5\frac{7}{8}$ feet of wood. How much wood did they have left after building the pillars?

____

5. Nina paints murals on buildings in her town. She used $5\frac{1}{2}$ gallons of red and green paint for her most recent mural. Nina used $1\frac{1}{2}$ gallons more red paint than green paint. How many gallons of each color did Nina use?

____

6. Frieda handed out 60 flags on three streets before the parade. Frieda handed out 26 flags on Main Street. If she handed out an equal number of flags on both Sycamore Drive and on Elm Street, how many flags did Frieda hand out to spectators on each of those two streets?

____

Name____

Classify Angles

Write *acute, right,* or *obtuse* for each angle.

1. $\angle AFD$ ______ **2.** $\angle BFD$ ______ **3.** $\angle DFE$ ______

4. $\angle CFD$ ______ **5.** $\angle CFE$ ______ **6.** $\angle AFB$ ______

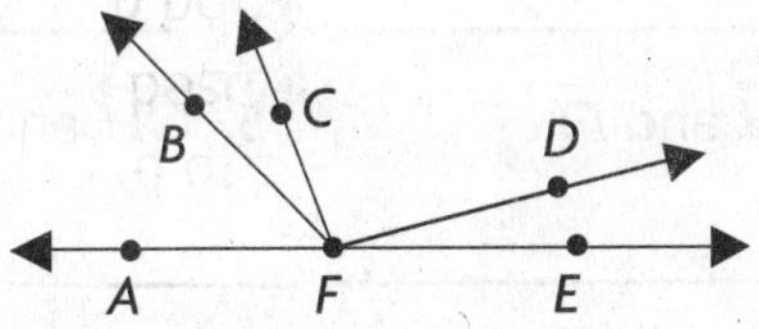

Write the numbers of acute, obtuse, and right angles.

7.

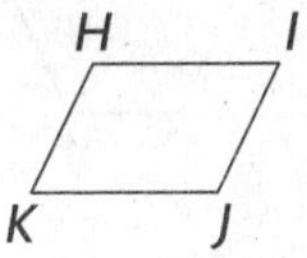

______
______

8.

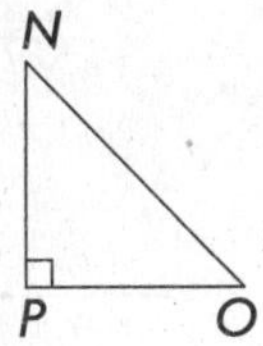

______
______

9.

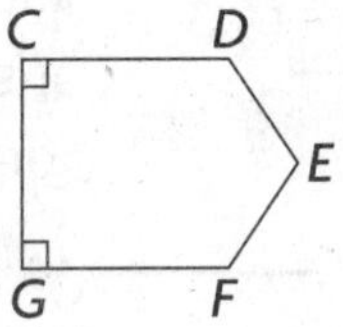

______
______

Problem Solving and TAKS Prep

10. The drawing at the right shows a triangle and a square. How many acute, right, and obtuse angles are used in the drawing?

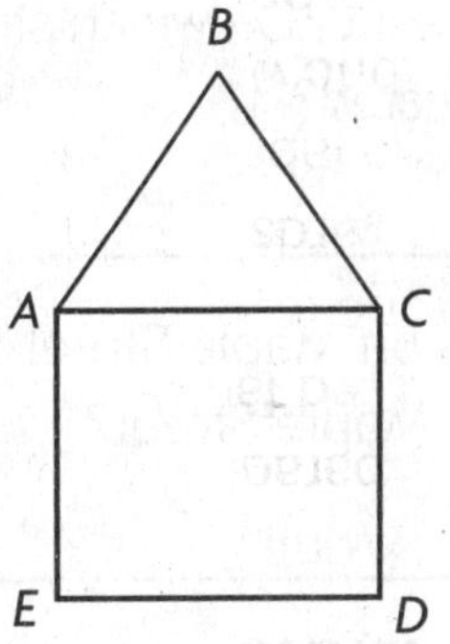

____

11. Which describes this figure?

A 2 acute angles
B 4 acute angles
C 2 right angles
D 4 right angles

12. Which describes this figure?

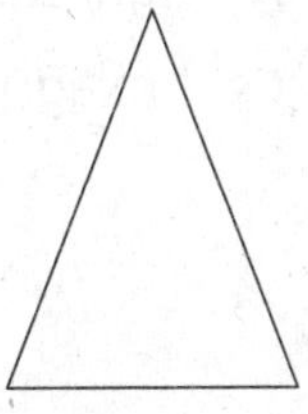

F no acute angles
G 3 acute angles
H 2 obtuse angles
J 3 obtuse angles

Line Relationships

Write *parallel, intersecting,* or *perpendicular* for each.

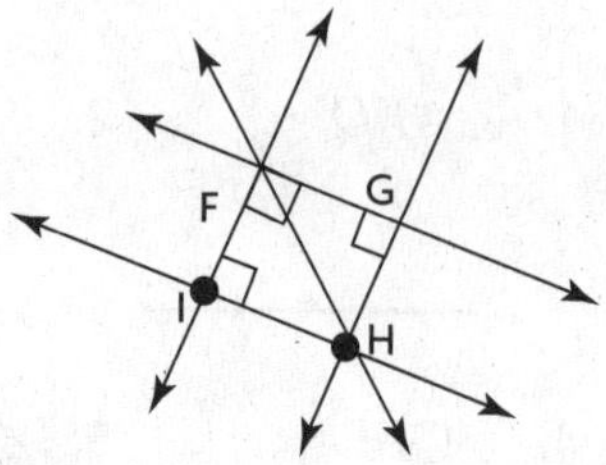

1. $\overleftrightarrow{FG}$ and $\overleftrightarrow{IH}$ ______________
2. $\overleftrightarrow{FH}$ and $\overleftrightarrow{FI}$ ______________
3. $\overleftrightarrow{GH}$ and $\overleftrightarrow{IH}$ ______________
4. $\overleftrightarrow{IH}$ and $\overleftrightarrow{FH}$ ______________
5. $\overleftrightarrow{GH}$ and $\overleftrightarrow{FG}$ ______________
6. $\overleftrightarrow{FI}$ and $\overleftrightarrow{GH}$ ______________

Identify the number of pairs of sides that are parallel and the number that are perpendicular.

7. 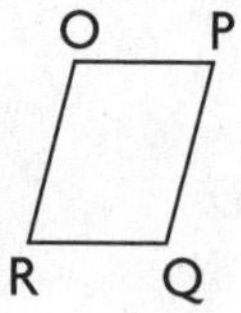

____
____

8. 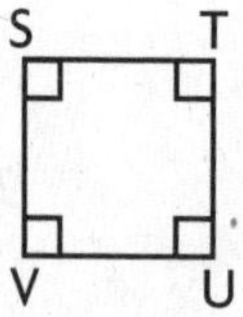

____
____

9. 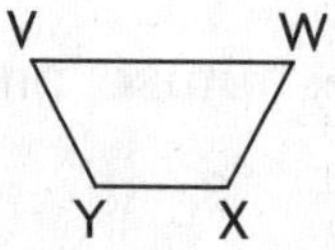

____
____

10. 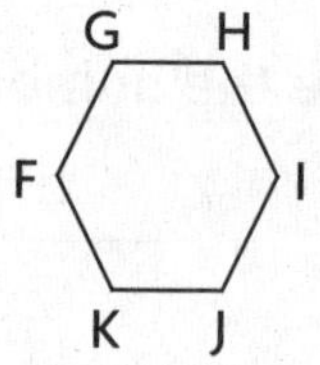

____
____

Problem Solving and TAKS Prep

USE DATA For 11–12, use the map.

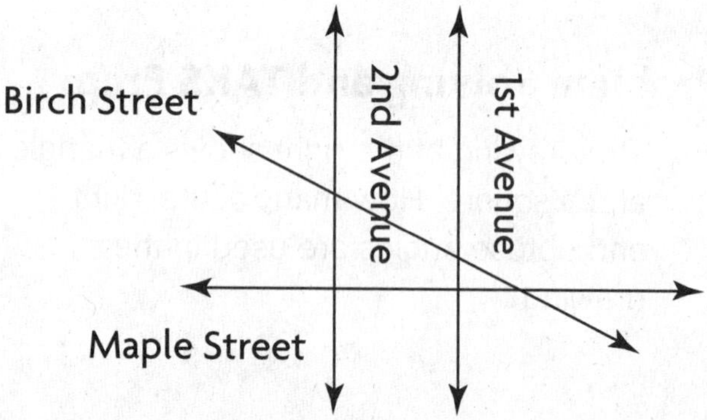

11. Maria's house is on a street that is parallel to 1st Avenue. On which street is Maria's house located?

____

12. Trisha's house is on Maple Street. What streets intersect Maple Street?

____

13. Which figure has 2 pairs of parallel lines?

   A right triangle **B** square
   C hexagon **D** pentagon

14. Which figure has 1 pair of perpendicular lines?

   F right triangle **G** square
   H pentagon **J** rectangle

Name________________________________

Polygons and Circles

Name each polygon, and tell whether it is regular or not regular.

1. 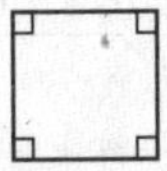

____

____

2. 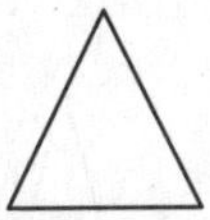

____

____

3. 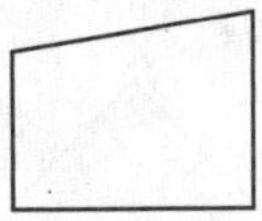

____

____

4. 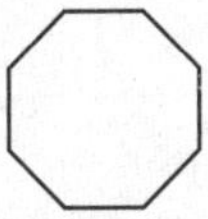

____

____

Complete 5–6. Then use a compass to draw each circle. Draw a radius and a diameter, and label their measurements.

5. radius = ☐

diameter = 1.4 cm

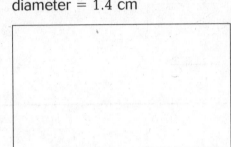

6. radius = 0.9 cm

diameter = ☐

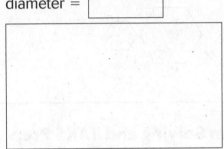

Problem Solving and TAKS Prep

USE DATA For 7–8, use the figures at the right.

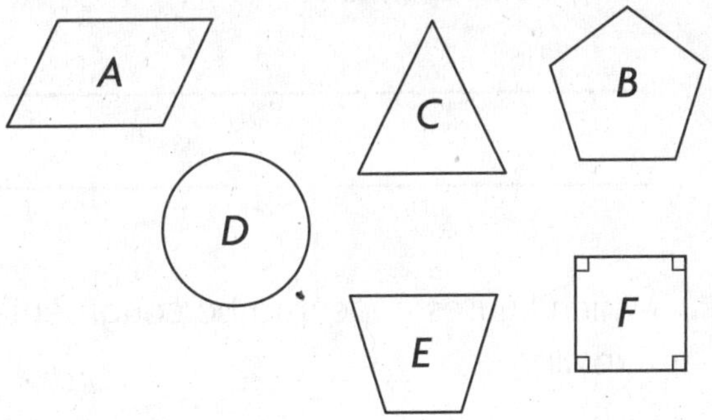

7. Which figures are polygons?

____

8. How does figure *A* compare to figure *F*? Explain how they are alike and how they are different.

____

____

9. Which best describes the figure?

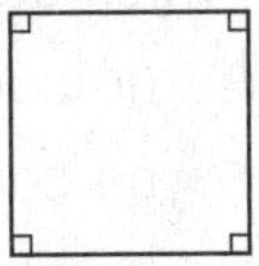

A square

B rectangle

C regular pentagon

D pentagon

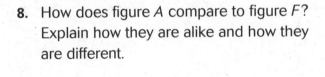

10. Which best describes the figure?

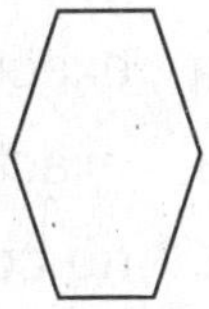

F regular pentagon

G pentagon

H regular hexagon

J hexagon

Name ______________________________ Lesson 12.4

Congruent Parts

For each figure, identify line segments and angles that are congruent.

1. 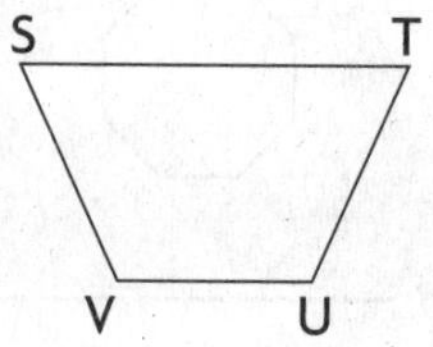

2. 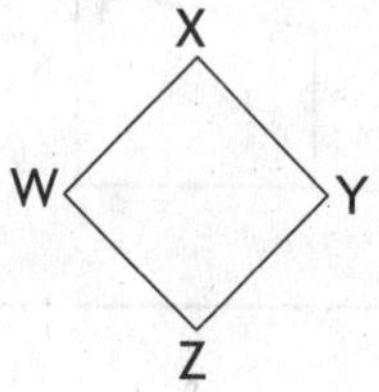

3. 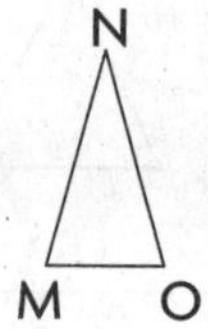

4. 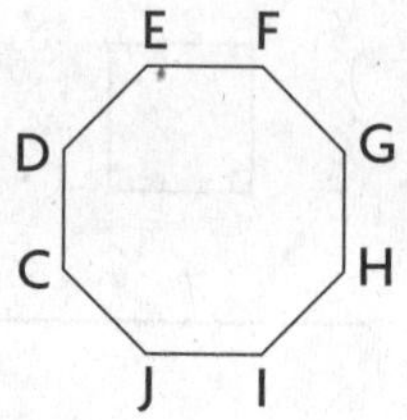

______ ______ ______ ______

Write whether the two figures appear to be *congruent* or *not congruent*.

5. 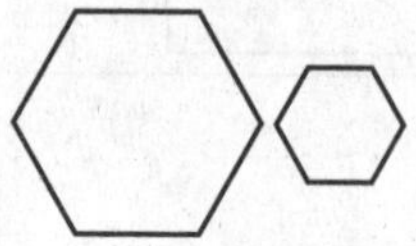

6. 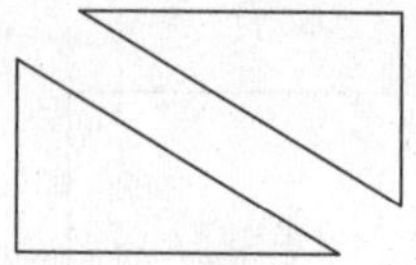

______ ______

Problem Solving and TAKS Prep

USE DATA For 7–8, use the figures shown.

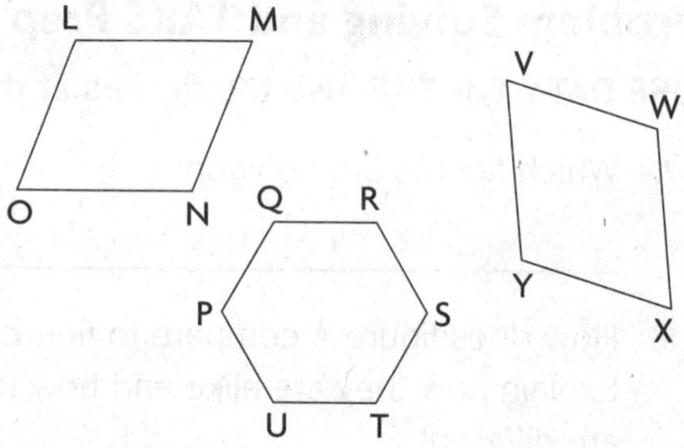

7. Which figure at the right is a regular polygon? Are the sides and angles in a regular polygon congruent? Explain.

______

______

8. Which figures appear to be congruent? Explain.

______

______

9. Which describes this figure?

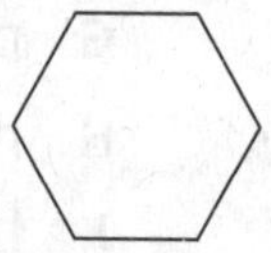

A exactly 6 congruent sides
B exactly 4 congruent sides
C no congruent angles
D no congruent sides

10. Which describes this figure?

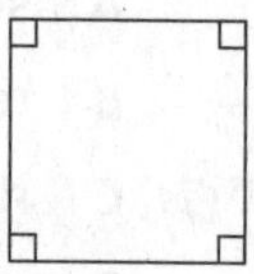

F no congruent sides
G no congruent angles
H exactly 2 congruent sides
J exactly 4 congruent sides

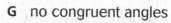

Name____ **Lesson 12.5**

Classify Triangles

Classify each triangle. Write *isosceles, scalene,* or *equilateral.* Then write *right, acute,* or *obtuse.*

1. 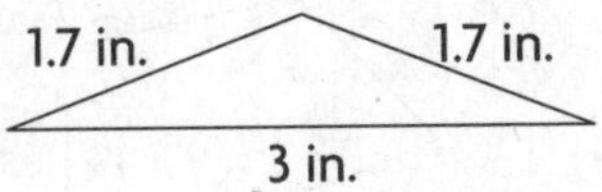

____

2. 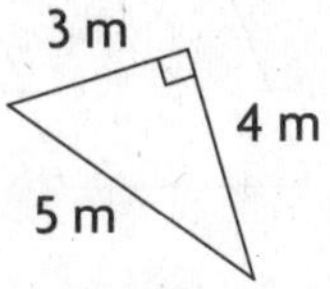

____

3. 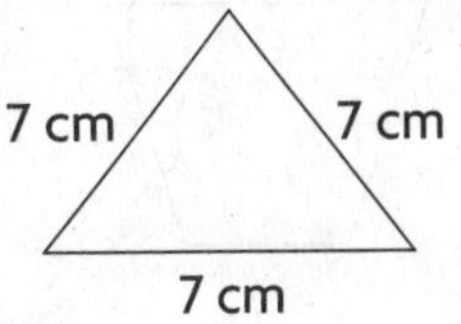

____

4. 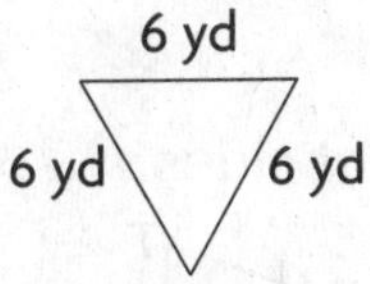

____

5. 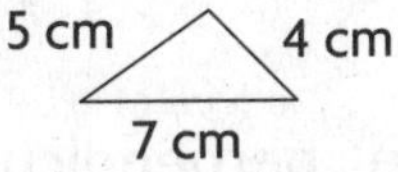

____

6. 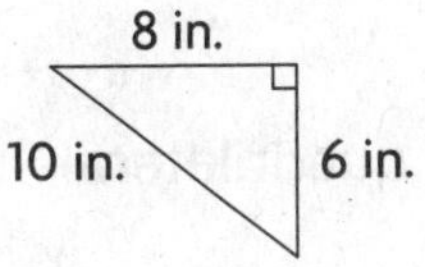

____

Classify each triangle by length of its sides or measure of its angles.

7. 2 ft, 5 ft, 4 ft ________________

8. 11 m, 11 m, 11 m ________________

9. 30°, 30°, 120° ________________

10. 30°, 70°, 80° ________________

11. 30°, 60°, 90° ________________

12. 1 cm, 3 cm, 3 cm ________________

Problem Solving and TAKS Prep

USE DATA For 13–14, use the drawing at the right.

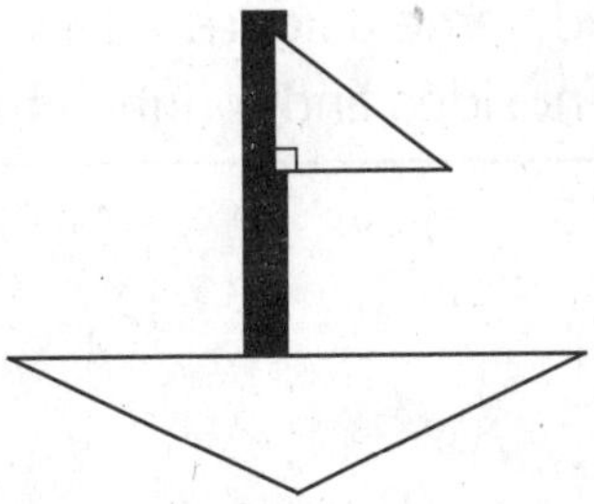

13. What type of triangle is the sail?

____

14. What type of triangle is the boat?

____

15. A triangle has three acute angles. What type of triangle is it?

A obtuse

B acute

C scalene

D right

16. A triangle has no congruent sides. What type of triangle is it?

F obtuse

G acute

H scalene

J equilateral

Name ________________________________

Classify Quadrilaterals

Classify each figure in as many ways as possible. Write *quadrilateral, parallelogram, square, rectangle, rhombus,* or *trapezoid.*

1. 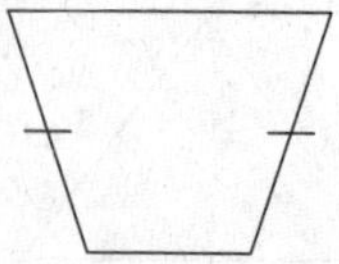

____

____

2. 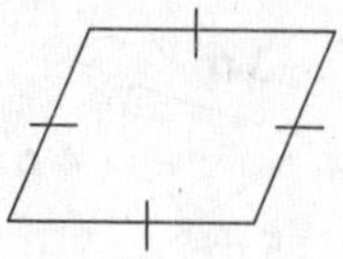

____

____

3. 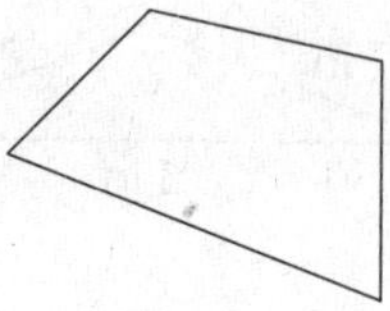

____

____

For each quadrilateral, name the parallel, perpendicular, and congruent sides.

4. 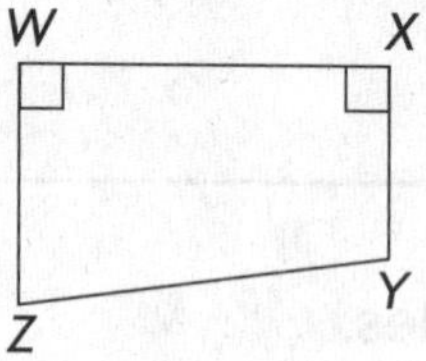

____

____

____

____

5. 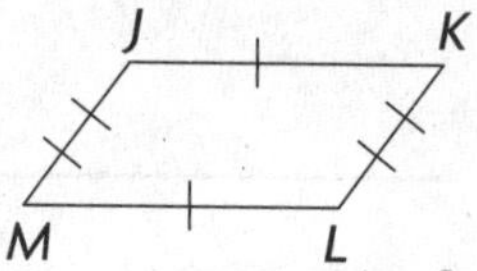

____

____

____

____

6. 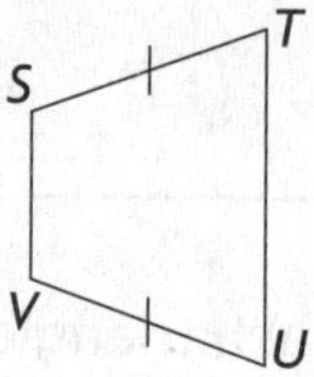

____

____

____

____

Problem Solving and TAKS Prep

7. Draw and name a figure with 4 congruent sides and 4 right angles.

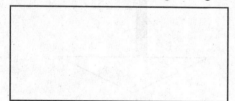

8. Draw and name a figure that has opposite sides parallel and congruent.

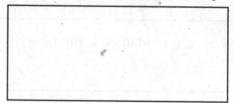

9. Which quadrilateral has exactly 1 pair of parallel sides?
   - **A** rectangle
   - **B** square
   - **C** trapezoid
   - **D** parallelogram

10. Which quadrilateral does NOT have opposite sides parallel and congruent?
   - **F** rectangle
   - **G** square
   - **H** trapezoid
   - **J** parallelogram

Name____

Symmetry

Draw all lines of symmetry. Then tell whether each figure has rotational symmetry by writing yes or no.

1. 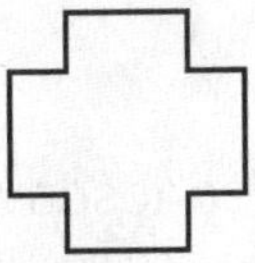

____

2. 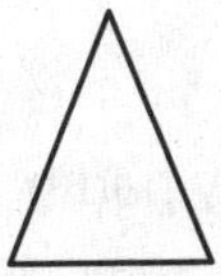

____

3. 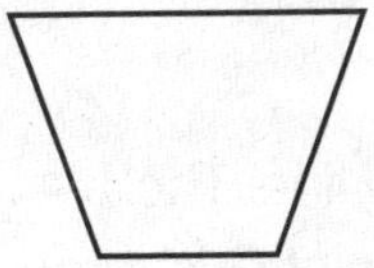

____

4. 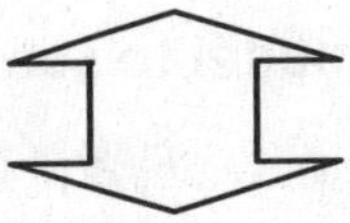

____

5. 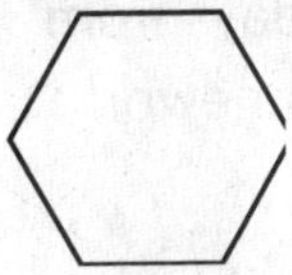

____

6. 

____

7. 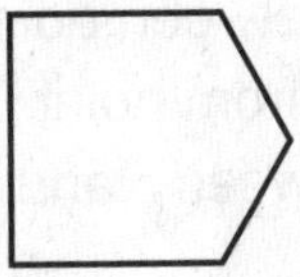

____

8. 

____

Each figure has rotational symmetry. Tell the fraction and the angle measure of the smallest turn that matches the original figure.

9. 

____

10. 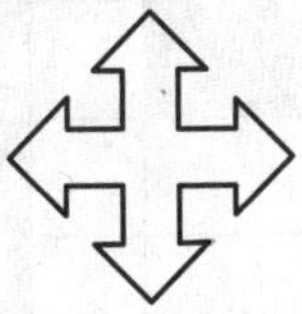

____

11. 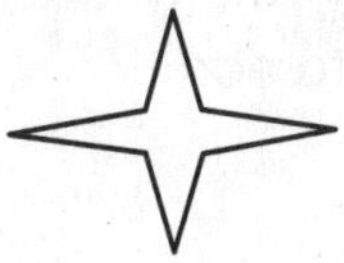

____

12. 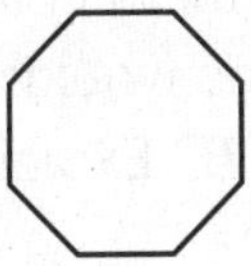

____

Problem Solving and TAKS Prep

13. Which figure has rotational symmetry?

A 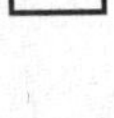

B

C

D

14. Which figure has rotational symmetry?

F 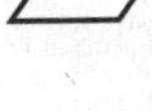

G

H

J

Name____

Problem Solving Workshop Strategy: Make a Model

Problem Solving Strategy Practice

Use a model to solve.

1. Points *E*, *F*, and *G* are on a line. The distance from point *F* to point *G* is three times the distance between points *E* and *F*. The distance between points *E* and *G* is four times the distance between points *E* and *F*. Point *E* is 8 inches from point *G*. How far is the distance between *F* and *G*?

____

2. Kelly had homework on Wednesday, Thursday, and Friday. Friday she had the least amount of homework. Wednesday she did not have the most homework. What is the order of these days from least homework to most homework?

____

Mixed Strategy Practice

USE DATA For 3–4, use the shaded figures shown.

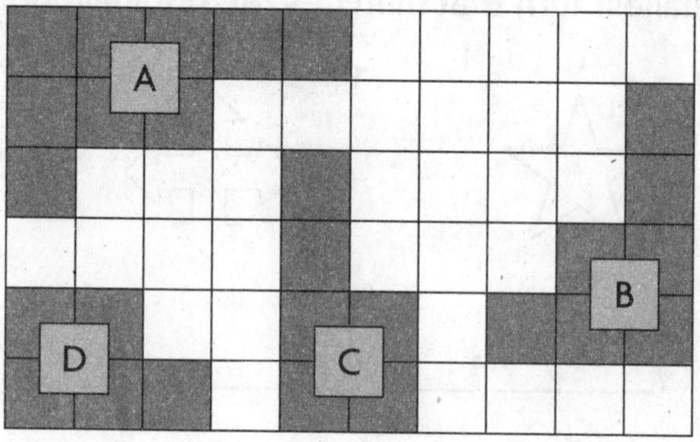

3. Compare figure *A* and figure *B*. To make them congruent how many squares would you need to add to figure *B*? Explain.

____

____

____

4. Which figure could be joined with figure *A* to form a rectangle?

____

5. Lucy builds a regular decagon out of toothpicks. She removes 4 toothpicks and builds a figure from the remaining toothpicks. Then she adds 2 toothpicks and builds another figure. Finally, she takes away all but 3 toothpicks. What three figures does Lucy build?

____

6. Nathan is going for a walk. He walks 3 blocks straight, then turns 90° to the right. He continues to walk for 4 blocks, then turns 45° to the right and walks 5 blocks. What figure does the path Nathan walks make?

____

Name____ **Lesson 13.1**

Three-Dimensional Figures

Identify each solid figure. Write *prism, pyramid, cylinder, cone,* or *sphere.*

1. 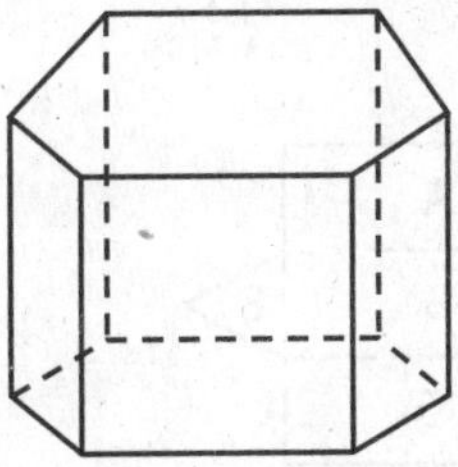

____

2. 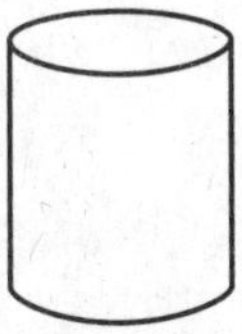

____

3. 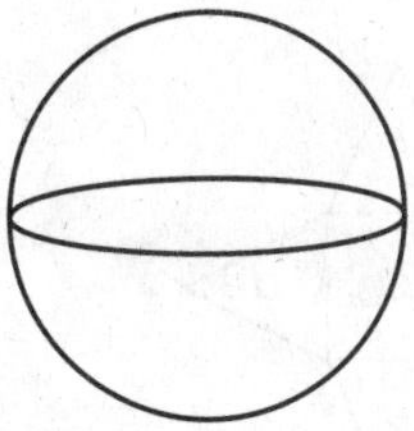

____

4. 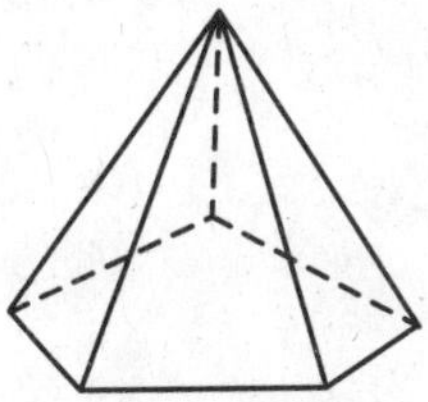

____

5. 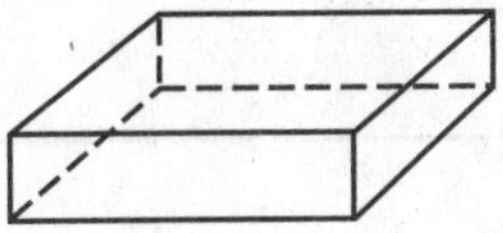

____

6. 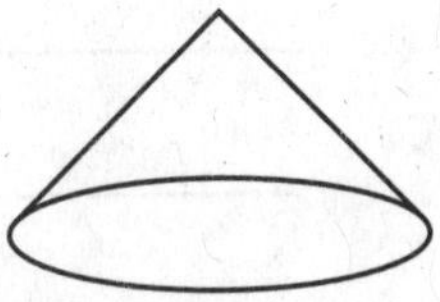

____

7. 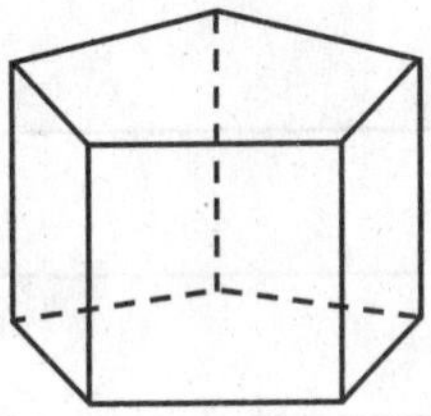

____

8. 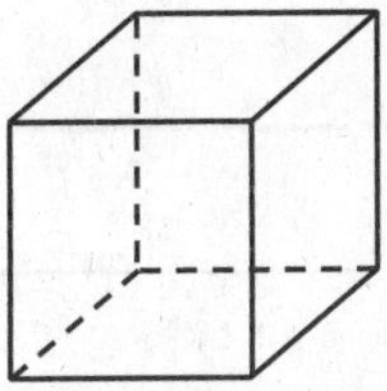

____

Name each three-dimensional figure described.

9. I have 0 bases and 1 curved surface.

____

10. I am a polyhedron with 2 congruent polygons as bases. Each of my bases has 6 sides.

____

Problem Solving and TAKS Prep

USE DATA For 11–12, use the pyramid at the right.

Great Pyramid in Egypt

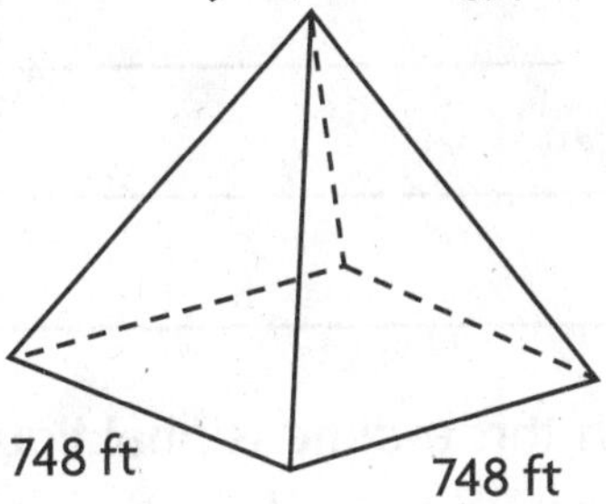

11. What is the shape of the base of the Great Pyramid in Egypt?

____

12. How many congruent faces does the Great Pyramid have?

____

13. Which solid figure has two bases that are congruent circles?

A cone

B cylinder

C sphere

D pentagonal pyramid

14. How many pairs of parallel faces does a rectangular prism have?

F 1 pair

G 2 pairs

H 3 pairs

J 4 pairs

Name________________________________

Model Three-Dimensional Figures

Identify the faces that are parallel, perpendicular, or congruent. Name each three-dimensional figure that can be made with the net.

1.

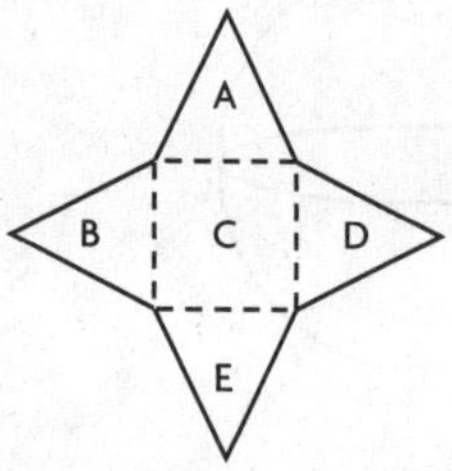

____

____

____

____

____

2.

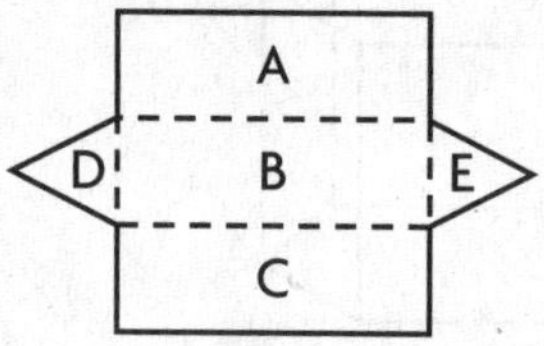

____

____

____

____

____

3.

____

____

____

____

____

4.

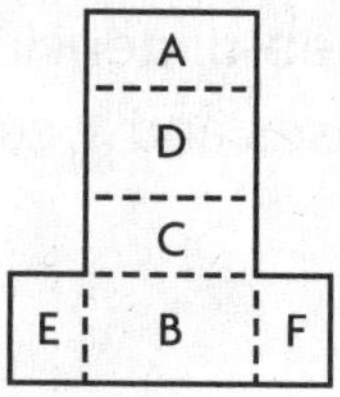

____

____

____

____

____

Name each three-dimensional figure.

5. This figure has 6 faces with 3 pairs of parallel faces, 4 faces that are perpendicular to each base, and 3 pairs of congruent faces.

____

6. This figure has 5 faces with 4 congruent faces, no parallel faces and no perpendicular faces.

____

7. The figure has 6 faces. All of the faces are congruent with 3 pairs of faces that are parallel.

____

Name ____________________

Identify Faces of Figures

For each figure, name the faces that are parallel, perpendicular, and congruent to the face shaded. All bases are regular polygons.

1.

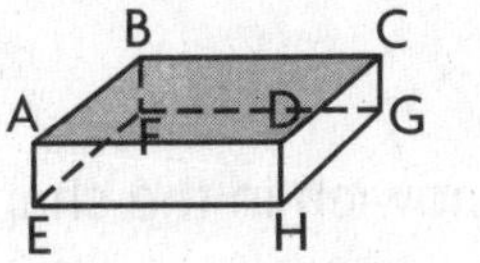

2.

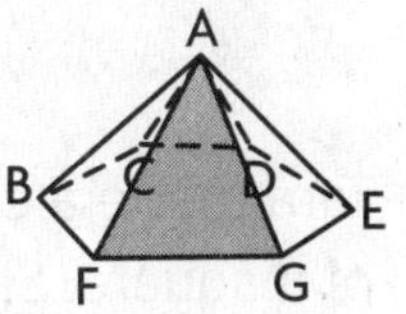

3.

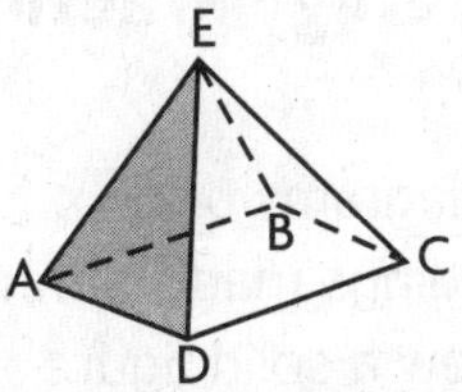

4.

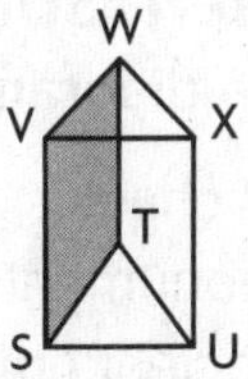

5.

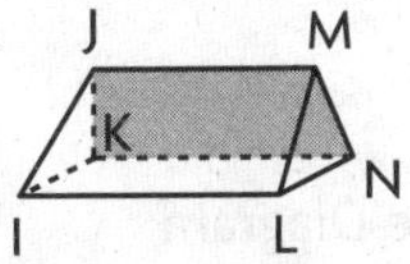

6.

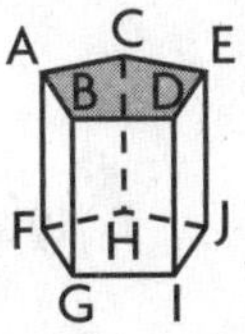

7.

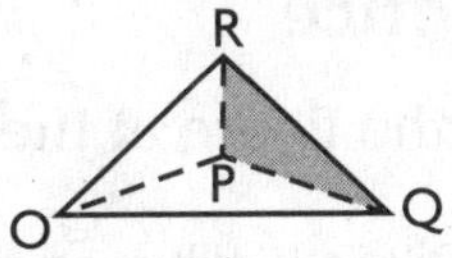

8.

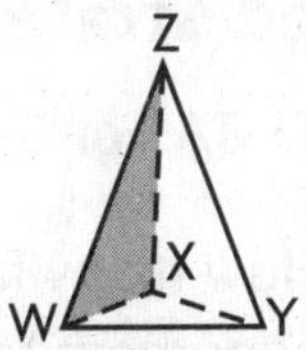

Problem Solving and TAKS Prep

9. Dalton is studying four of his little sister's toy blocks. He has a cube block, a square pyramid block, and a triangular prism block. Which block has no parallel or perpendicular sides?

____

10. Jamie has a block with all congruent faces, but the block has no parallel or perpendicular faces. Which solid figure is Jamie's block?

____

11. How many parallel faces does a triangular prism have?

A 6

B 4

C 2

D 0

12. How many perpendicular faces does a pentagonal prism have?

F 10

G 5

H 2

J 1

Name________________________________ Lesson 13.4

Problem Solving Workshop Strategy: Draw a Diagram

Problem Solving Strategy Practice

Draw a diagram to solve.

1. Michael is drawing diagrams of monuments and buildings made during ancient times. He drew a solid figure with 4 congruent, triangle faces and a base with 4 equal sides. What solid figure did Michael draw?

____

2. Tara made a club sandwich in the shape of a square. She then cut the sandwich into pieces, using all possible lines of symmetry. If Tara wants use a toothpick to hold each piece of the sandwich together, how many toothpicks will she need?

____

Mixed Strategy Practice

USE DATA For 3–4, use the figure at the right.

Billy's Birdhouse Diagram

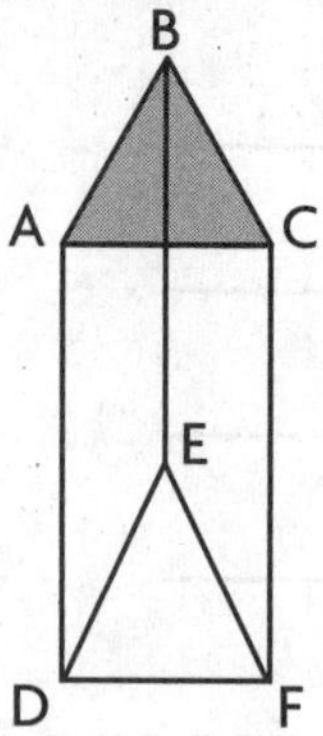

3. How many faces of Billy's birdhouse are congruent to face *ABED*? Name the faces and identify the shape of each face.

____

____

____

4. Billy drew the diagram shown as a sketch for a birdhouse he wants to build. Which face is parallel to the shaded face of Billy's diagram?

____

5. Lucia makes candles. She wants one candle to be a solid figure with two congruent, parallel bases with 5 sides. What solid figure will the candle be?

____

6. Peter, Ricardo, Diana, and Jules all play tennis. Each plays 3 matches with each of the others. What is the total number of tennis matches they play?

____

Name________________________ **Lesson 14.1**

Algebra: Graph Ordered Pairs

For 1–6, use the coordinate grid to name the ordered pair for each point.

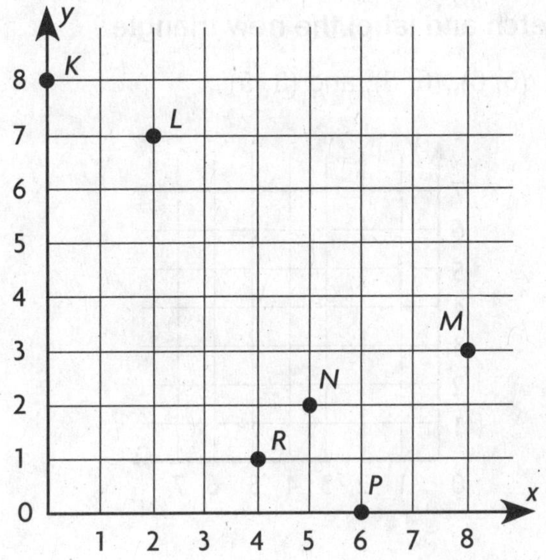

1. *R* ________ **2.** *P* ________ **3.** *M* ________

4. *L* ________ **5.** *N* ________ **6.** *K* ________

Graph and label the following points on the coordinate grid.

7. *S* (3,3) **8.** *T* (0,2) **9.** *U* (4,6)

10. *W* (7,5) **11.** *X* (8,4) **12.** *Y* (4,0)

13. *B* (8,1) **14.** *D* (2,5) **15.** *Q* (7,7)

Problem Solving and TAKS Prep

16. Start at (0,0). Move 6 units to the right and 4 units up. What ordered pair is the ending point?

____

17. Start at the origin. Move 5 units to the right and 3 units up. Then move 2 more units to the right and 2 more units up. What ordered pair is the ending point?

____

18. The point (3,0) is:

A not an ordered pair
B on the *x*-axis
C on the origin
D on the *y*-axis

19. The point (0,0) is:

F not an ordered pair
G on the *x*-axis
H on the origin
J on the *y*-axis

Name______________________________________

Translations on a Coordinate Grid

Graph a triangle with the given vertices. Then translate each vertex of the triangle 3 units down and 3 units left. Sketch and label the new triangle.

1. (6, 6), (6, 5), and (3, 5)

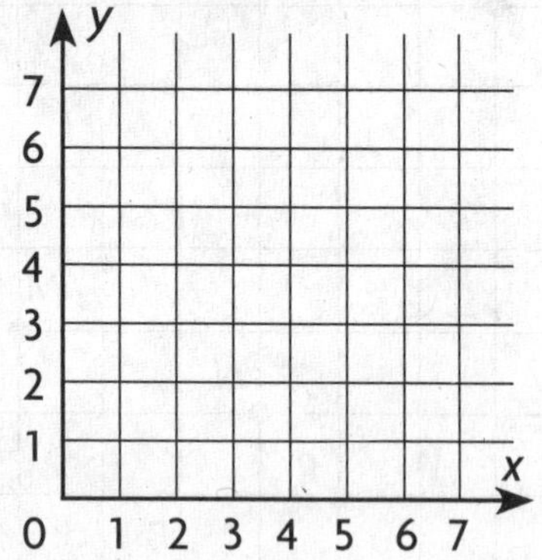

2. (4, 7), (3, 5), and (3, 7)

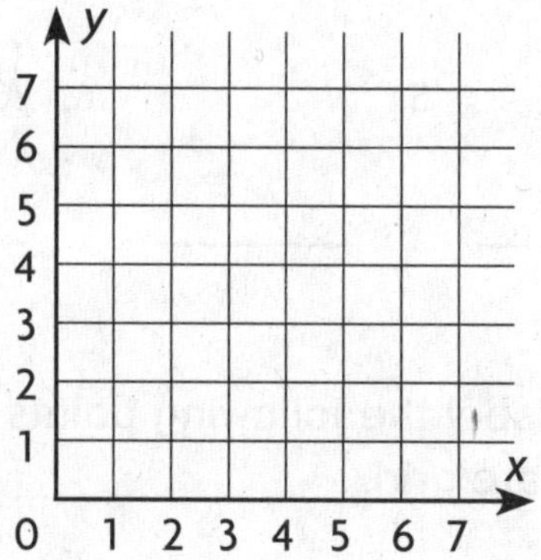

Problem Solving and TAKS Prep

USE DATA For 3-4, use the grid at the right.

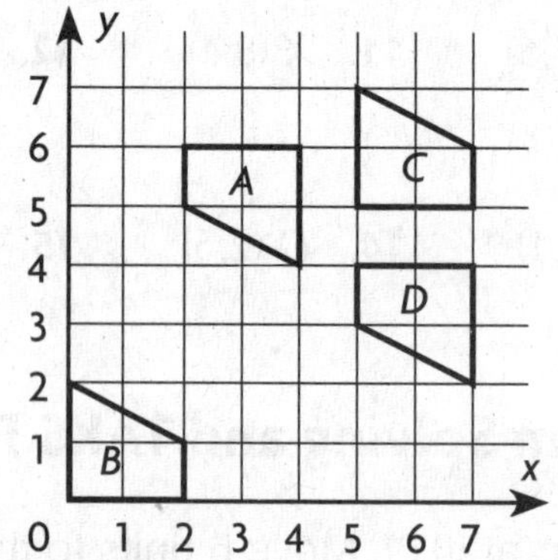

3. Tell which Quadrilateral could be the result of the translation of Quadrilateral *A*.

____

4. Tell which Quadrilateral could be the result of the translation of Quadrilateral *B*.

____

5. A triangle with vertices (2, 1), (0, 1), and (1, 3) will translate 4 units up. Which shows the ordered pairs for the vertices of the new triangle?

A (6, 1), (4, 1), (5, 3)

B (2, 5), (0, 5), (1, 7)

C (6, 5), (4, 5), (5, 7)

D (5, 2), (5, 0), (7, 1)

6. A triangle with vertices (1, 1), (1, 3), and (5, 1) will translate 3 units right. Which shows the ordered pairs for the vertices of the new triangle?

F (1, 1), (1, 3), (5, 1)

G (1, 4), (1, 7), (5, 4)

H (4, 4), (4, 7), (8, 4)

J (4, 1), (4, 3), (8, 1)

Name____

Reflections on a Coordinate Grid

Graph a triangle with the given vertices. Draw a horizontal line of reflection. Then reflect the triangle over the line. Sketch and label the new triangle.

1. (4, 6), (1, 5), and (3, 4)

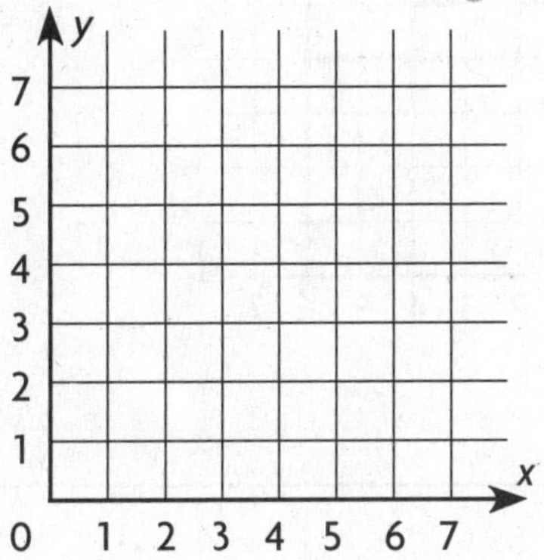

2. (0, 3), (1, 1), and (2, 3)

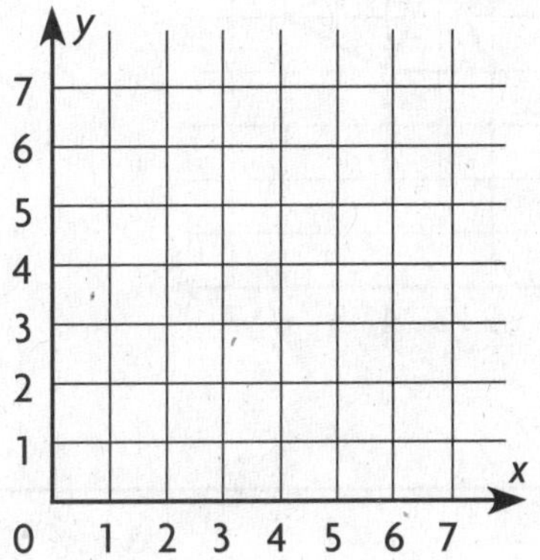

Problem Solving and TAKS Prep

USE DATA For 3 and 4, use the grid at the right.

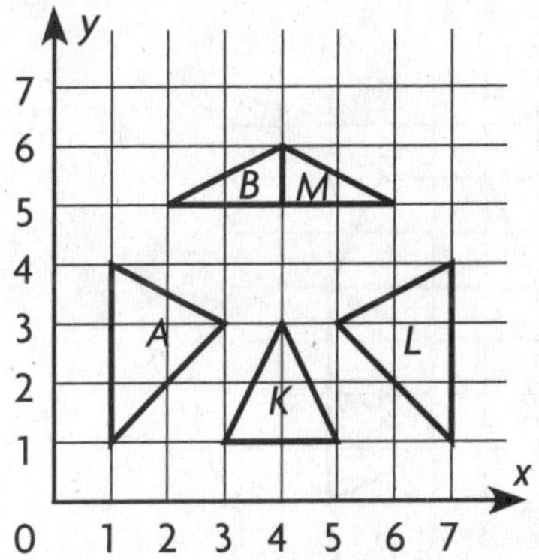

3. Tell which figure could be the result of the reflection of Triangle *A*.

____

4. Tell which figure could be the result of the reflection of Triangle *B*.

____

5. In a reflection over a horizontal line, which coordinates will change?

A only *x*
B only *y*
C both *x* and *y*
D neither *x* nor *y*

6. In a reflection over a vertical line which coordinates will change?

F both *x* and *y*
G only *x*
H neither *x* nor *y*
J only *y*

Name ____________________ Lesson 14.4

Rotations on a Coordinate Grid

Tell the type of rotation shown of the figure on the left. Write *clockwise* or *counterclockwise,* and write $\frac{1}{4}$, $\frac{1}{2}$, or $\frac{3}{4}$.

1.

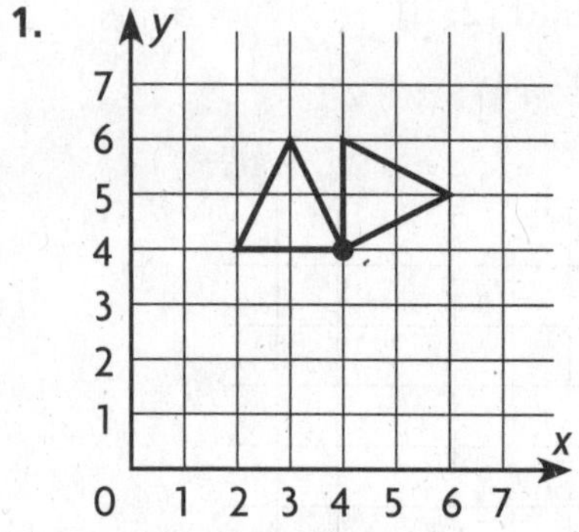

2.

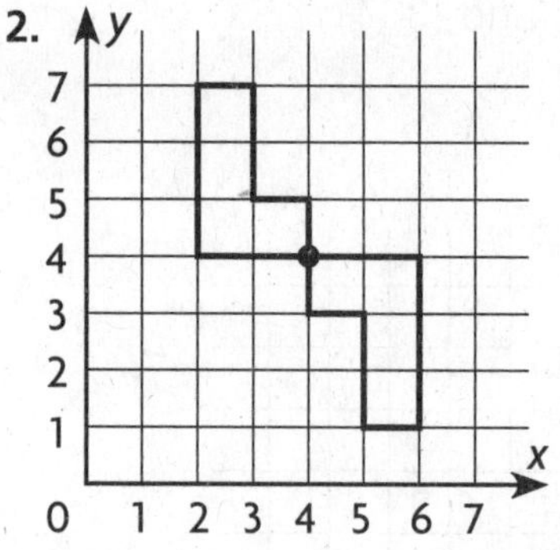

____

____

Problem Solving and TAKS Prep

For 3-4, tell how the first figure was moved. Write translation, reflection, or rotation. For a rotation, write clockwise or counterclockwise, and write $\frac{1}{4}$, $\frac{1}{2}$, or $\frac{3}{4}$.

3.

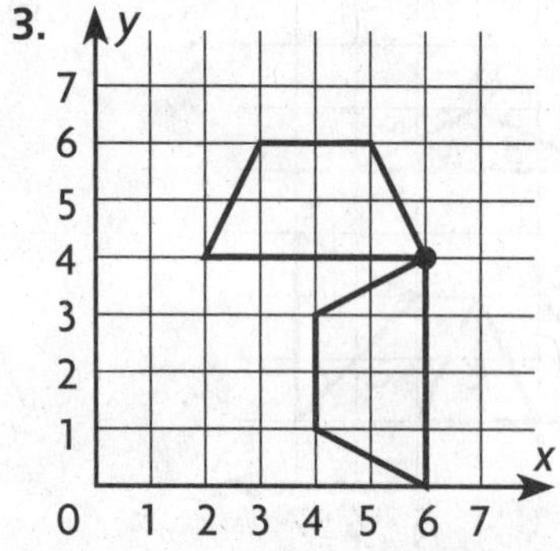

4.

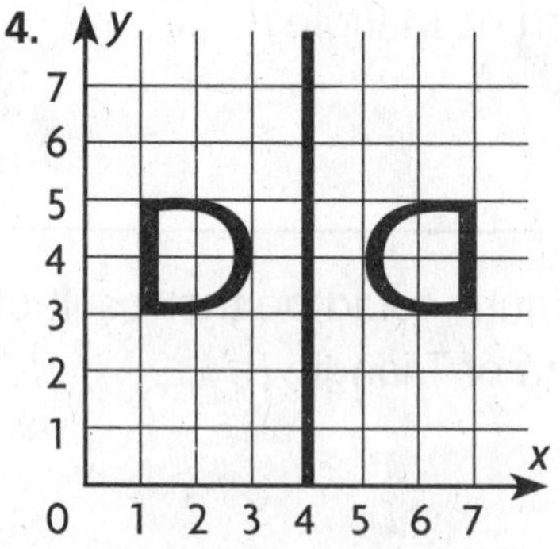

____

____

5. In a rotation, which coordinates may change?

A only *x*

B only *y*

C both *x* and *y*

D neither *x* nor *y*

6. Which type of transformation describes the movement of a figure to a new position along a straight path?

F reflection

G translation

H transformation

J rotation

Name______________________________

Tessellations

Predict whether or not the figures will tessellate. Trace and cut out several of each figure. Tell if the figure, or pair of figures, will tessellate. Write *yes*, or *no*.

1.

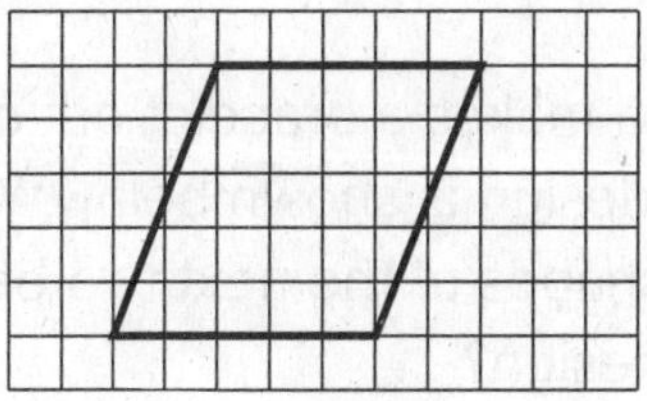

____

2.

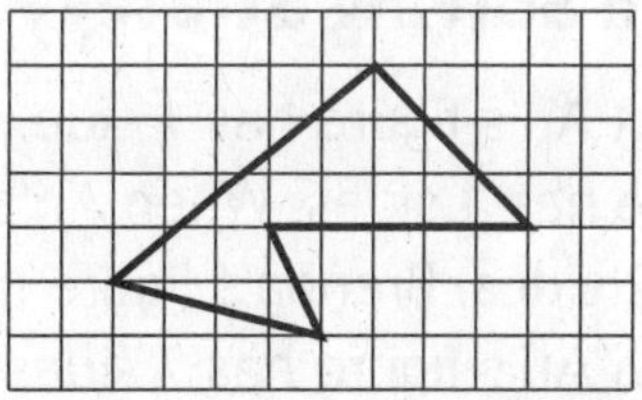

____

3.

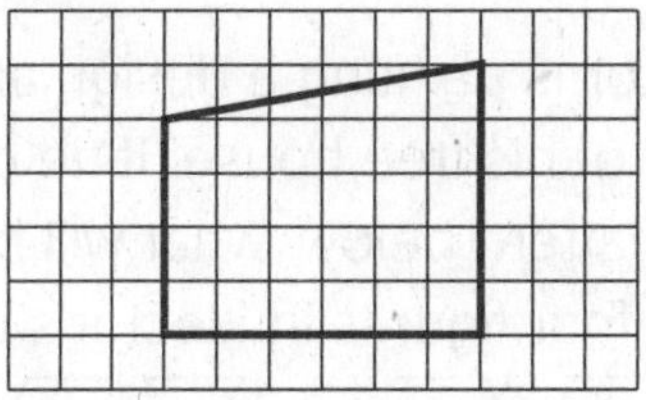

____

4.

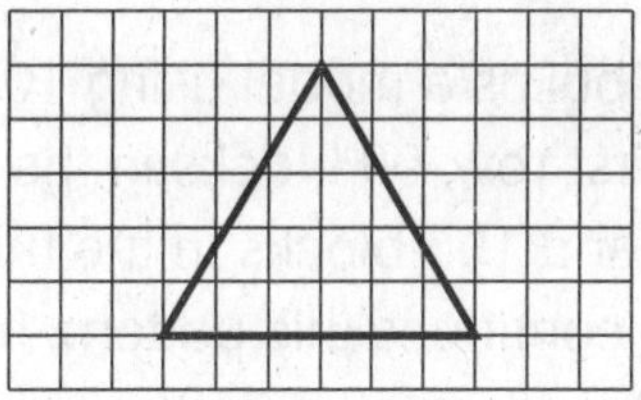

____

5.

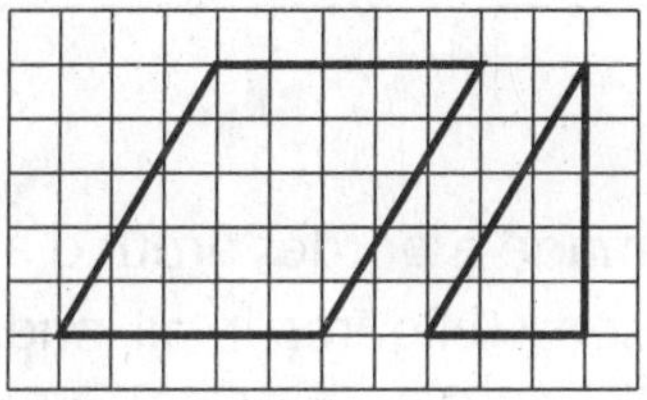

____

6.

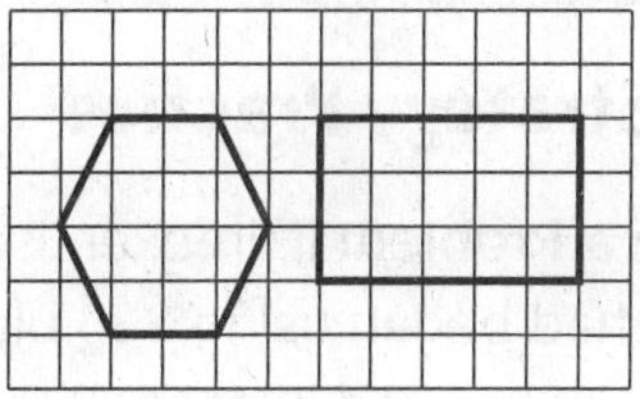

____

7.

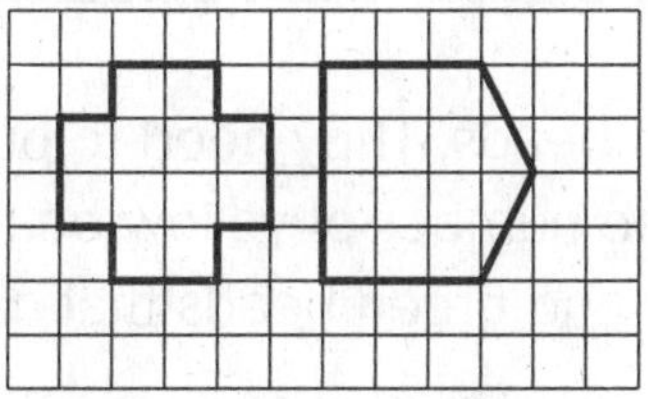

____

8.

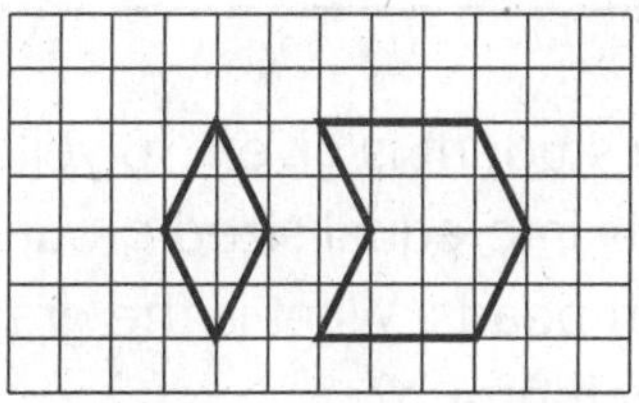

____

Name ______________________________

Problem Solving Workshop Strategy: Look for a Pattern

Problem Solving Strategy Practice

1. When Ari's figure has 1 side, Brenda's figure has 4 sides. When Ari's figure has 2 sides, Brenda's figure has 6 sides. When Ari's figure has 7 sides, how many sides does Brenda's figure have?

____

2. Tonya makes a bracelet out of beads. Her design is shown below. What are the shapes of the next two beads in the design?

____

3. Julia builds a model using 105 blocks in the first row, 90 blocks in the second row, and 105 blocks in the third row. If Julia continues this pattern, how many blocks will she use in the fourth row?

____

4. Hector is painting a design around the floor of his tree house. If he continues the pattern below, what will be the next four figures in Hector's design?

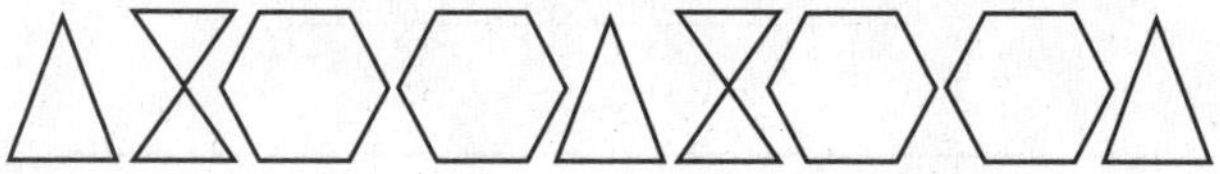

____

Mixed Strategy Practice

5. **Pose a Problem** If, in exercise 3 above, Julia had began using 230 blocks in the first row but had kept the same pattern then how many blocks would there be in the fourth row? Show your work.

____

6. Rose made a border around a painting. She used 40 figures in all, and used her pattern unit 8 times. How many figures are in Rose's pattern unit?

____

7. Each student is given 36 yellow beads and 32 green beads. They need to put the beads into equal sized groups, each having the same number of yellow beads and green beads. What is the greatest number of yellow and green beads that can be in each group?

____

Name________________________________ Lesson 15.1

Number Patterns

Write a rule for each pattern. Then find the missing number.

1. 1, 4, 7, 10, ____
____

2. 2, 8, 14, ____, 26
____

3. 4, 8, 16, 32, ____
____

4. 2, 8, 32, ■
____

5. 3, 8, 23, ■
____

6. 5, 11, 23, ■
____

7. 80, 40, 20, ■
____

8. 1, 4, 16, ■
____

9. 6, 14, 30, ■
____

Problem Solving and TAKS Prep

10. Tess wants to purchase a new bicycle. She wants to save the same amount each time she gets paid. She begins with $15. Find a rule and fill in the table of Tess's savings.

Month	1	2	3	4	5	6
Amount per month	$15	$29	$43	$57	■	■

____

11. Carlos collects comic books. He buys the same number of new comic books each month. Find a rule and fill in the table to find out how many new comic books Carlos buys each month.

Month	1	2	3	4	5	6
Comic books per month	3	9	15	21	■	■

____

12. Which pattern is described by this rule: Multiply by 2, add 1?

A 6, 13, 27, 45

B 9, 18, 36, 64

C 4, 9, 19, 39

D 2, 6, 14, 30

13. Which pattern is described by this rule: Multiply by 3, subtract 1?

F 1, 2, 5, 14

G 2, 6, 18, 54

H 4, 11, 33, 98

J 3, 5, 15, 45

Name____ **Lesson 15.2**

Geometric Patterns

Write a rule for the pattern. What might be the next figure in the pattern?

1. 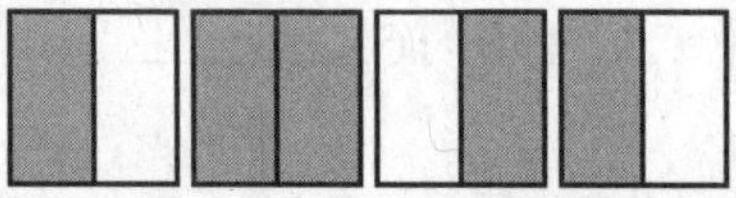

____

____

2. 

____

____

3. 

____

____

4. 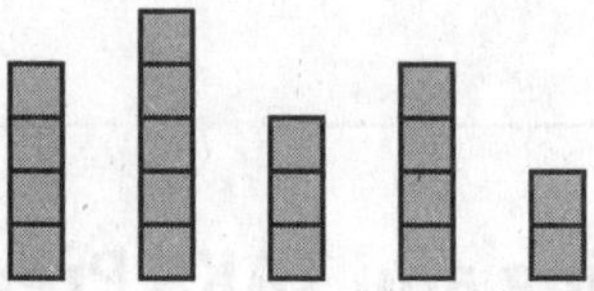

____

____

Problem Solving and TAKS Prep

5. Tobias will use the pattern below to make a picture frame border. Write a rule for the pattern. What are the next two pieces of the pattern?

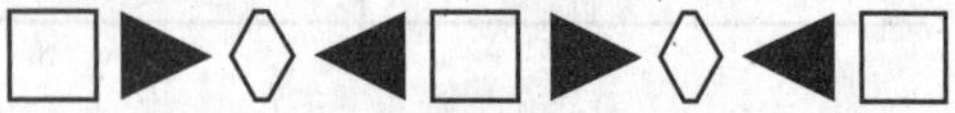

6. Ileana will use this rule for a pattern: Start with 1 dot in 1 square. Add 1 dot to each square to make the next figure. What could be the first three figures of Ileana's pattern?

7. What is the next figure in the pattern?

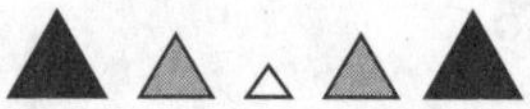

A 

B 

C 

D 

8. What is the next figure in the pattern?

F 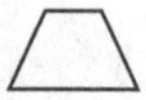

G 

H 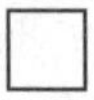

J 

Name________________________ **Lesson 15.3**

Problem Solving Workshop Skill: Make Generalizations

Problem Solving Skill Practice

Make generalizations to solve the problems.

1. Rolando uses the boxes below to make a pattern for multiples of 6. 6×1 is shown by 1 box with 6 dots inside, 6×2 is shown by 2 boxes each with 6 dots inside and so forth. If Rolando's pattern continues, how many boxes of 6 dots will he show for 6×9?

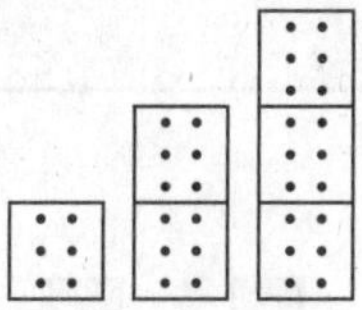

____

2. ABC Farms and Super Yum applesauce are packaged in jars of the same size and shape. ABC Farms brand contains 32 ounces of applesauce in each jar. How many ounces of applesauce are there in 8 jars of Super Yum brand?

____

3. Bright & Bold and Super Shiny are two brands of paint. They are packaged in tins of the same size and shape. If Bright & Bold paint is sold in 64 ounce tins, how many ounces of paint are there in 12 tins of Super Shiny brand?

____

4. Stephanie needs 30 yards of thread to finish her quilt. If World's Best thread spools each contain 5 yards, how many yards of thread are there in 12 spools of World's Best? How many spools will Stephanie need to buy to finish her quilt?

____

Mixed Applications

USE DATA For 5–7, use the table.

Monty's Comic Book Fund

Month	Amount saved
1	$15
2	$30
3	$45
4	$60

5. Monty saves his money to add to his comic book collection. How much money will he have in his fund at the end of 9 months? Explain the generalization you used to solve the problem.

____

____

6. If Monty decides to spend $5 of his comic book fund each month, how much will have saved after 9 months?

____

7. Gisella saves $\frac{1}{2}$ as much as Monty each month. How much money will Gisella have at the end of 8 months?

____

Create a Pattern

Tell how each pattern might have been created.

1.

2.

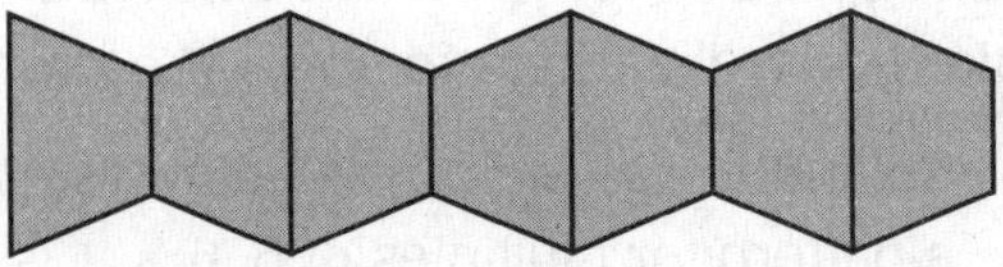

3.

4.

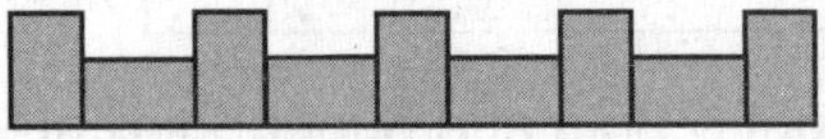

Trace each figure. Then transform it to create a pattern. Sketch your design.

5. Translate the figure horizontally four times.

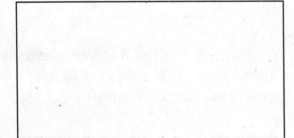

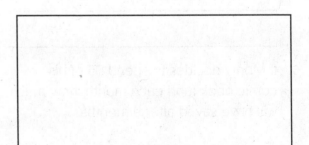

6. Draw a point of rotation. Rotate the figure clockwise $\frac{1}{4}$ turn five times.

Name____

Relationships: Coordinate Grid

Complete each table. Write the ordered pairs. Then graph them.

1.

x	1	2	3	4	5
y	2	3	4	5	■

____

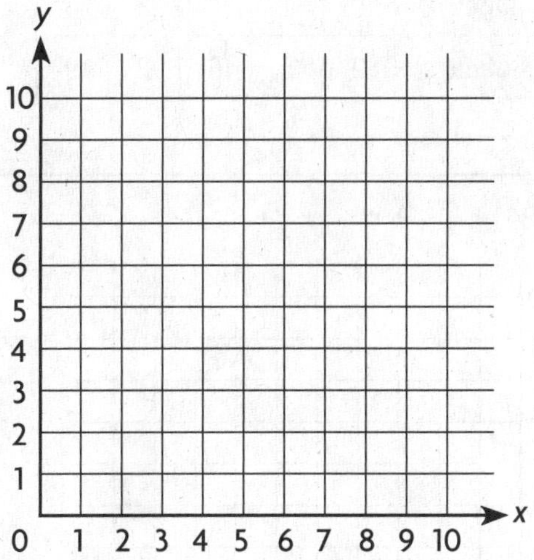

2.

x	0	1	2	3	4
y	1	3	5	7	■

____

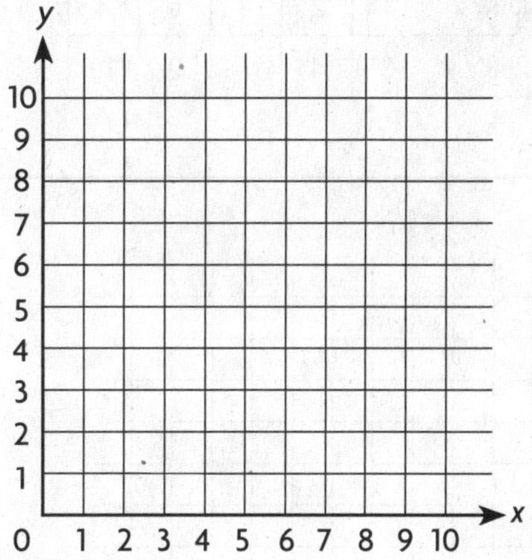

3.

x	0	2	4	6	8
y	2	4	6	8	■

____

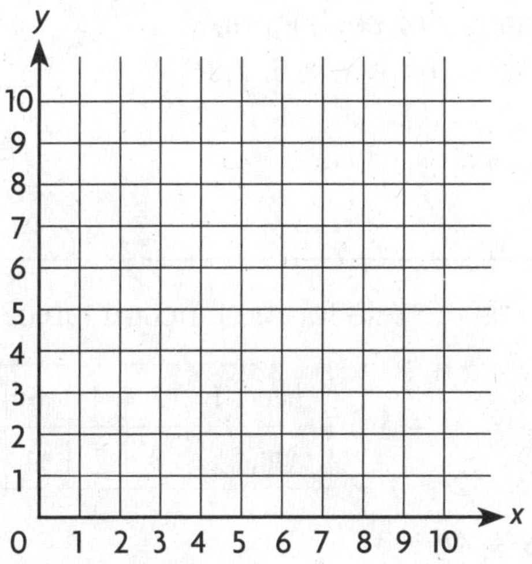

4.

x	2	4	6	8	10
y	1	3	5	7	■

____

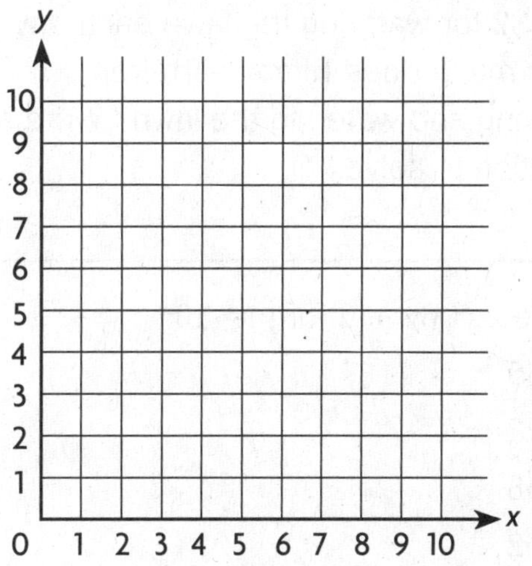

Name____

Functions

Find a rule to complete the table. Then write the rule as an equation.

1.

input, *t*	2	5	7	9	12
output, *p*	5	8	10	12	

____

2.

input, *a*	3	5	7	9	11
output, *b*	6	10	14	18	

____

Write the ordered pairs. Then graph them.

3.

input, *x*	1	2	4	6	8
output, *y*	3	4	7	9	10

____

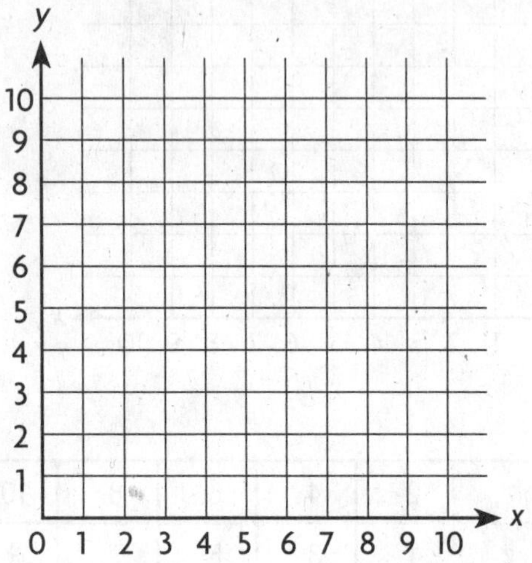

4.

input, *x*	0	1	2	3	4
output, *y*	3	5	6	8	9

____

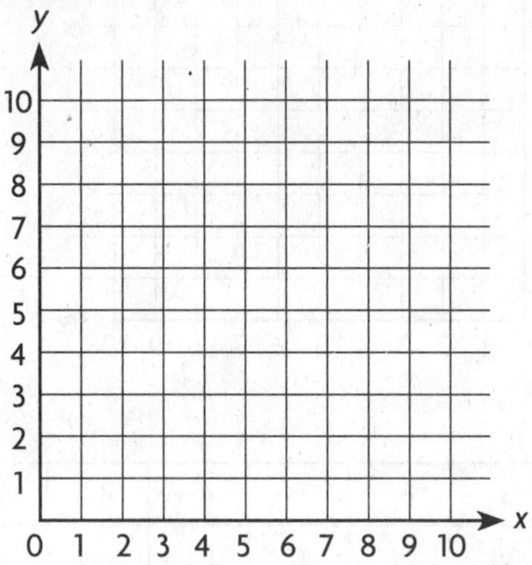

Problem Solving and TAKS Prep

5. Tiffany charges $15 for mowing a lawn and $2 for watering the lawn each day. How much does Tiffany earn from mowing and watering the lawn? Write a function to solve.

____

6. **Algebra** Use the function. Find the output, y, for each input, x.
$y = 4x - 6$ for $x = 5, 7, 9$

____

7. Solve $x = 6y + 2$ for $y = 10$.

A 65

B 22

C 58

D 62

8. What is the rule for the function table?

F $s = t + 3$

G $s = 3t + 1$

H $s = 2t - 1$

J $s = 4t + 3$

input, *t*	2	4	6	8
output, *s*	3	7	11	15

Name______________________________

Relationships and Graphs

Write a rule as an equation to show the function. Then graph the ordered pairs.

1.

t	1	2	3	4	5
p	3	6	9	12	15

____

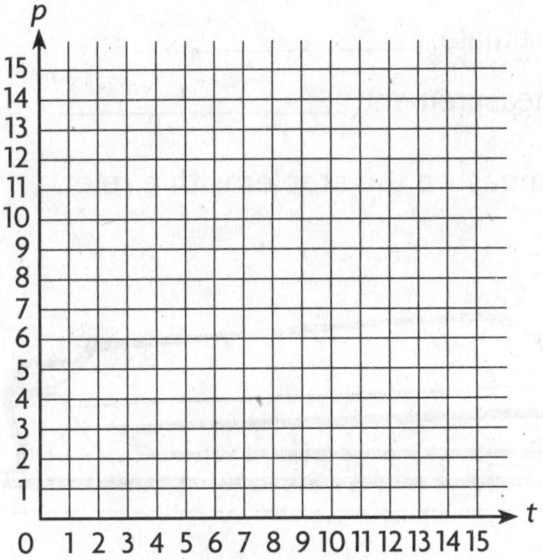

2.

a	12	10	8	6	4
b	6	5	4	3	2

____

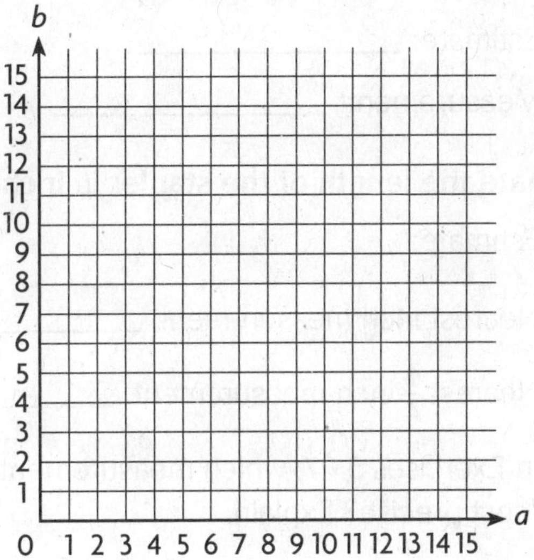

Problem Solving and TAKS Prep

3. Write a rule as an equation to show the function for the table below.

x	8	12	16	20	24
y	4	6	8	10	12

____

4. Solve $y = 4x + 7$ for $x = 7, 9, 10,$ and 11.

____

5. Which is not an ordered pair for the table below?

k	5	10	15	20	25
m	1	2	3	4	5

A (5, 1)

B (25, 5)

C (20, 4)

D (3, 15)

6. Which is not an ordered pair for the table below?

f	0	2	4	6	8
g	4	8	16	24	32

F (2, 8)

G (6, 24)

H (4, 0)

J (8, 32)

Name______________________________ **Lesson 16.1**

Customary Length

Estimate the length of four items or distances of your choice in inches, feet or yards. Then measure the item or distance and record your actual measurement.

1. Item or distance: ______________

   Estimate: ______________

   Measurement: ______________

2. Item or distance: ______________

   Estimate: ______________

   Measurement: ______________

3. Item or distance: ______________

   Estimate: ______________

   Measurement: ______________

4. Item or distance: ______________

   Estimate: ______________

   Measurement: ______________

Estimate the length of the stapler in inches. Then measure the stapler with a ruler.

5. Estimate: ______________

6. Nearest inch measurement: __________

7. Nearest $\frac{1}{4}$-inch measurement: ________

8. In Exercises 5–7, which measurement is most precise? Explain.

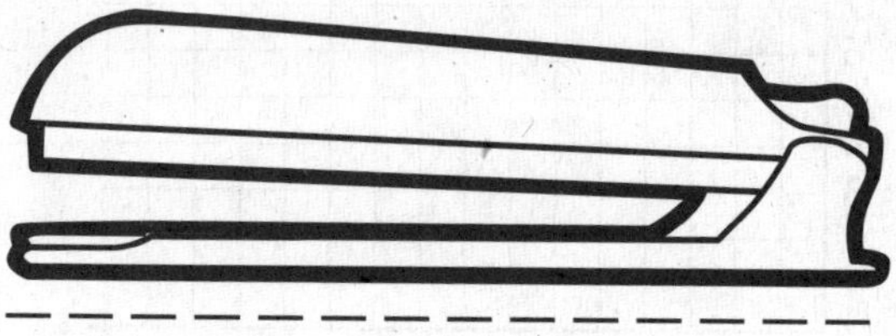

____

____

Circle which measurement is more precise.

9. $4\frac{1}{8}$ inches or $4\frac{1}{4}$ inches

10. 1 foot or $11\frac{1}{2}$ inches

11. $\frac{7}{8}$ inches or $\frac{3}{4}$ inches

Estimate the length in inches. Then measure to the nearest $\frac{1}{8}$ inch.

12. 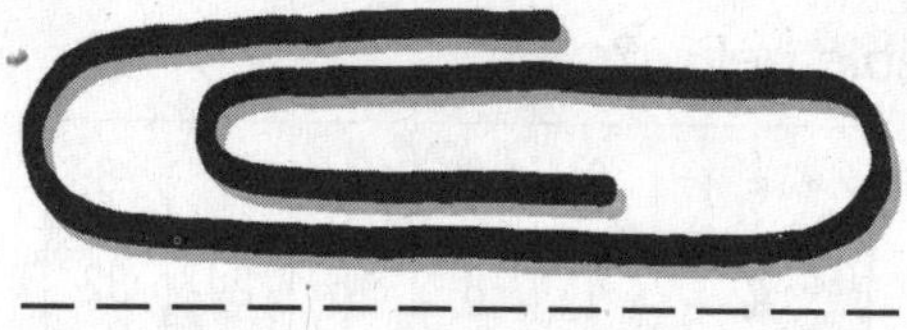

   Estimate: ______________

   Measurement: ______________

13. 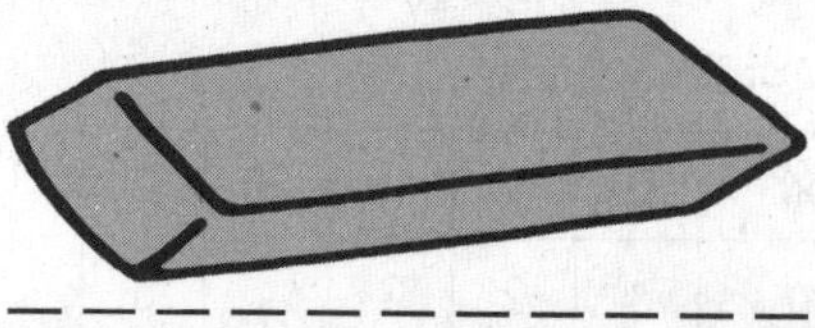

   Estimate: ______________

   Measurement: ______________

Name______________________________ **Lesson 16.2**

Choose Customary Units

Circle the unit you would use to measure each.

1. distance from home plate to first base
inches or yards

2. width of your textbook
inches or feet

3. height of a lamp post
feet or miles

4. weight of a calculator
ounces or pounds

5. amount of sugar for one batch of cookies
cups or gallons

6. weight of a dump truck
ounces or tons

7. weight of a chair
ounces or pounds

8. capacity of a car's fuel tank
gallons or quarts

9. distance from Miami to Atlanta
miles or feet

Choose the tool you would use to measure each.

10. height of your teacher ______________

11. weight of a steak ______________

12. capacity of a water bottle ______________

13. length of a caterpillar ______________

14. width of a cell phone ______________

15. length of a pool ______________

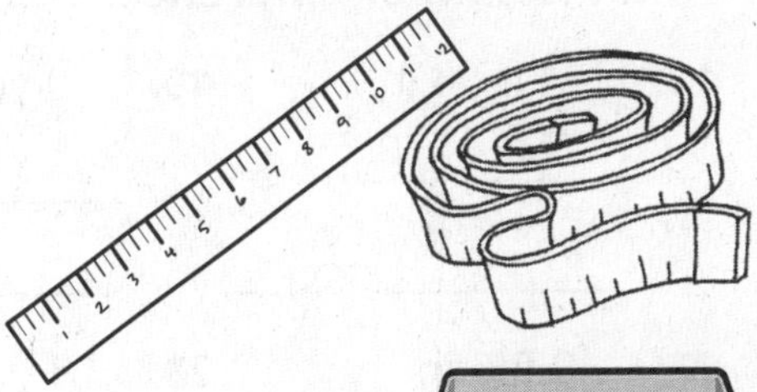

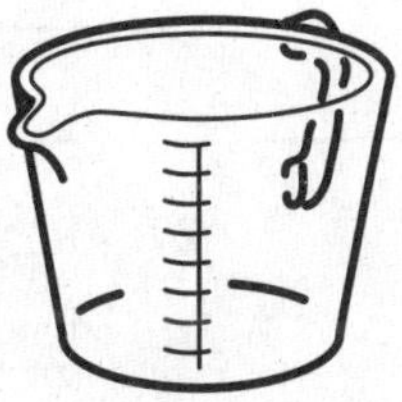

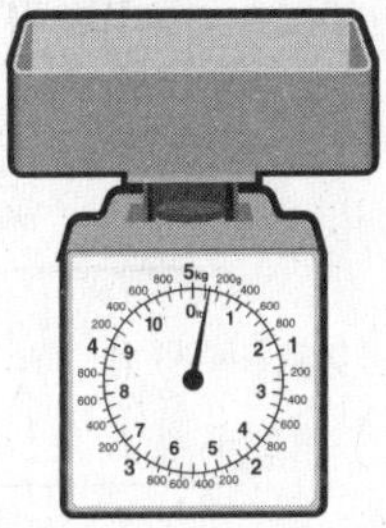

Problem Solving and TAKS Prep

16. Quin needs to measure the width and length of his bedroom. Which tool and units should he use?

____

17. Carmelita is making a casserole that requires one cup of sliced mushrooms. Which measuring tool should she use?

____

18. Which unit would you use to measure the weight of an apple?

A ounces
B tons
C inches
D gallons

19. Which tool would you use to measure flour for a batch of waffles?

F ruler
G scale
H thermometer
J measuring cup

Name____

Change Customary Linear Units

Change the unit.

1. 12 yd = ■ ft
2. 24 in. = ■ ft
3. 9 ft = ■ in.
4. 1 yd 9 in. = ■ in.
5. 2 mi = ■ yd
6. 10 ft 11 in. = ■ in.
7. 12 ft = ■ yd
8. 144 in. = ■ ft
9. 2 yd 2 ft = ■ in.

Complete.

10. 10 ft 12 in. = ■ yd 2 ft
11. 2 mi 100 ft = 1 mi ■ ft
12. 2 yd 4 ft = ■ yd 1 ft
13. 5 ft 24 in. = ■ yd 1 ft
14. 120 in. = ■ yd 1 ft
15. 5 ft 2 in. = 4 ft ■ in.
16. 12 yd 1 ft = 10 yd ■ ft
17. 3 yd 1 in. = 2 yd ■ in.

Find the sum or difference.

18. 3 ft 6 in. + 2 ft 11 in. = ________
19. 9 yd 1 ft − 1 yd 2 ft = ________
20. 13 ft 7 in. + 2 ft 7 in. = ________
21. 10 yd 2 ft − 1 yd 1 ft = ________
22. 8 ft 6 in. + 4 ft 8 in. = ________
23. 3 yd 11 in. + 1 yd 2 in. = ________
24. 4 ft 4 in. − 8 in. = ________
25. 3 ft 6 in. − 2 ft 10 in. = ________
26. 9 ft + 4 ft 11 in. = ________
27. 2 yd 15 in. + 27 in. = ________
28. 12 ft 2 in. − 2 ft 8 in. = ________
29. 4 ft 1 in. − 1 ft 9 in. = ________

Problem Solving and TAKS Prep

30. The width of Jade's garden is 3 yd 1 ft. How many inches wide is the garden?

____

31. Peri glued two pieces of wood together. One was 9 in long; the other was 1 ft 3 in. How long is the glued piece?

____

32. Stu cut 4 inches off a 5-ft piece of rope. How long is the rope that is left?

A 5 ft 4 in.
B 4 ft 8 in.
C 4 ft 10 in.
D 4 ft 4 in.

33. A flagpole 7 yd 1 ft tall is the same height as which of the following?

F 17 ft 1 in.
G 21 ft
H 22 ft
J 24 ft 6 in.

Name______________________________

Customary Capacity and Weight

Change the unit.

1. 5 lb = ■ oz
2. 16 c = ■ qt
3. 8 gal = ■ qt
4. 4,500 lb = ■ T
5. 72 oz = ■ lb
6. 12 fl oz = ■ c
7. 16 qt = ■ gal
8. 10 c = ■ qt
9. 4.5 lb = ■ oz

Find the sum or difference.

10. 7 lb 6 oz + 4 lb 10 oz = ______
11. 11 gal 2 c − 2 gal 1 c = ______
12. 4 pt 1 c + 1 pt 1 c = ______
13. 23 lb 2 oz − 20 lb 14 oz = ______
14. 2 c 2 fl oz + 4 c 6 fl oz = ______
15. 3 qt 3 c + 4 qt 2 c = ______
16. 2 T 200 lb − 1 T 20 lb = ______
17. 4 pt 2 fl oz − 2 pt 6 fl oz = ______

Find the missing measurement.

18. 1 c + ■ = 2 qt
19. 12 fl oz + ■ = 2 c
20. 33 oz + ■ = 4 lb
21. 4 pt + ■ = 4 gal
22. 2 c + ■ = 1 gal
23. 1,500 lb + ■ = 1 T
24. 2 fl oz + ■ = 1 pt
25. 8 oz + ■ = 3.5 lb

Problem Solving and TAKS Prep

26. Mrs. Moore handed out 4 ounces of almonds to each of her 22 students. How many pounds of almonds did she hand out?

____

27. Camryn made 3 gallons of iced tea for a birthday party. How many cups of iced tea did she make?

____

28. Tito's baby sister weighed 128 ounces when she was born. How many pounds did she weigh?

A 12 lb
B 7 lb
C 8.5 lb
D 8 lb

29. Riley drank 8 cups of water during a soccer tournament. How many fluid ounces did he drink?

F 64 fl oz
G 32 fl oz
H 16 fl oz
J 64 qt

Name________________________________

Elapsed Time

Write the time for each.

1. Start: 7:14 A.M.
Elapsed time: 2 hr 50 min
End: ______________

2. Start: ______________
Elapsed time: 12 hr 3 min
End: 6:57 P.M.

3. Start: 4:12 P.M.
Elapsed time: ______________
End: 6:43 P.M.

4. Start: January 1, 3:00 A.M.
Elapsed time: 4 days 3 hr 30 min
End: ______________

5. Start: ______________
Elapsed time: 22 hr 12 min
End: 11:12 P.M.

6. Start: Monday, 2 P.M.
Elapsed time: ______________
End: Tuesday, 6 A.M.

Add or subtract.

7. 3 days 2 hr + 1 day 10 hr = ______________

8. 12 min 22 sec − 2 min 32 sec = ______________

9. 2 hr 12 min + 1 hr 49 min = ______________

10. 6 wk 6 days − 4 wk 5 days = ______________

11. 32 min 9 sec + 40 min 10 sec = ______________

12. 6 hr 6 min − 4 hr 19 min = ______________

13. 1 day 12 hr + 2 days 14 hr = ______________

14. 5 wk 3 days − 4 wk 6 days = ______________

Problem Solving and TAKS Prep

15. Christian checked out a book from the library that is due in 2 weeks. If he checks it out on April 3, what is the due date?

April

Sun	Mon	Tue	Wed	Thu	Fri	Sat
			1	2	3	4
5	6	7	8	9	10	11
12	13	14	15	16	17	18
19	20	21	22	23	24	25
26	27	28	29	30		

____

16. Mr. Lee requests that Ava and her classmates read for 25 minutes at home each weekday. How much time will they spend reading at home over 3 weeks?

____

17. Josh swam every Monday and Friday in June. How many days did he swim?

A 4 days

B 6 days

C 8 days

D 10 days

June

Sun	Mon	Tue	Wed	Thu	Fri	Sat
		1	2	3	4	5
6	7	8	9	10	11	12
13	14	15	16	17	18	19
20	21	22	23	24	25	26
27	28	29	30			

18. The movie started at 7:10 P.M. If it ended at 9:04 P.M., how long did it last?

F 2 hours, 6 minutes

G 1 hour, 54 minutes

H 1 hour, 56 minutes

J 2 hours, 4 minutes

Name______________________________________ **Lesson 16.6**

Problem Solving Workshop Strategy: Make a Table

Problem Solving Strategy Practice

Make a table to solve.

1. Sol is taking the bus to the skate park. He leaves home at 9 A.M. and walks for 10 minutes to the bus stop. He needs to be home by 3 P.M. Fill in the table below the bus schedule. How much time can he spend at the skate park?

____

Bus Schedule	
Depart	**Arrive**
Sol'Bus Stop	**Skate Park**
9 A.M.	9:15 A.M.
9:20 A.M.	9:35 A.M.
9:40 A.M.	9:55 A.M.
Skate Park	**Sol's Bus Stop**
2:15 P.M.	2:30 P.M.
2:35 P.M.	2:50 P.M.
2:55 P.M.	3:10 P.M.

2. Sol is planning to take the 2:35 bus home, but it is 10 min late. Does that make Sol late getting home? Explain.

____

____

Part of Trip	Start	End
Home to bus stop	9:00 A.M.	■
Bus to park	■	■
Skate park	■	■
Park to bus stop	■	■
Bus stop to home	■	■

Mixed Strategy Practice

USE DATA For 3–4, use the table.

3. Emma has just started at a new school where each class or activity lasts 50 minutes. Use the table on the right to complete her schedule.

4. How long is Emma's new school day?

____

Class or Activity	Start	End
Math	8:30 A.M.	■
Music	■	■
Spanish	■	11:00 A.M.
Lunch, recess	■	■
Novels	■	■
P.E.	12:40 P.M.	■
Science, Tech	■	■
Social Studies	■	3:10 P.M.

5. In a skate park competition, the first place competitor had 150 points. The combined points of the second and third place competitors was 50 points more than the winner. The third place contestant had 10 points less than second place, and the fourth place score was 20 points less than second place. What was each competitor's score? Complete the table on the right.

Competition Results	
First Place	150 points
Second Place	■
Third Place	■
Fourth Place	■

Name____________________

Lesson 17.1

Choose SI (Metric) Units

Choose the unit you would use to measure each.

1. length of a tennis court kilometers or meters

2. capacity of a mug milliliters or kiloliters

3. height of a stop sign millimeters or meters

4. mass of a digital camera kilograms or grams

5. capacity of a juice pitcher milliliters or liters

6. mass of a pick-up truck kilograms or milligrams

7. length of a post card centimeters or meters

8. mass of a container of strawberries grams or kilograms

9. distance from Chicago to Detroit kilometers or meters

Choose the tool you would use to measure each.

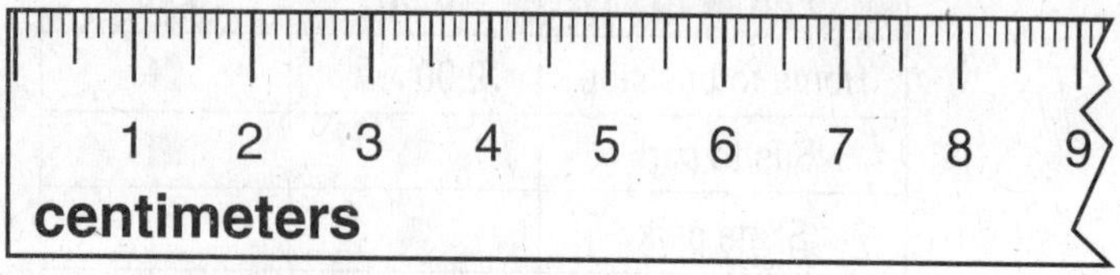

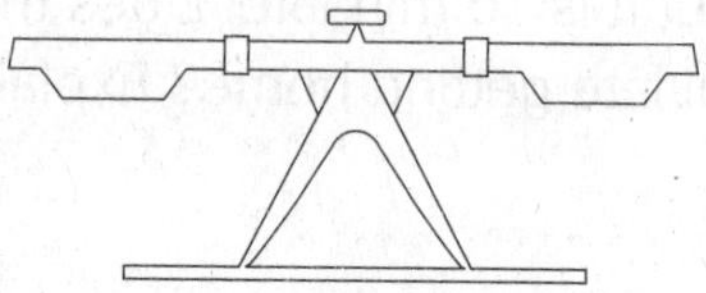

10. capacity of a soup bowl

11. mass of a pair of scissors

12. capacity of a water glass

13. mass of a paper clip

14. mass of a stuffed animal

15. width of a drawer

16. thickness of a phone book

17. mass of a baseball

Problem Solving and TAKS Prep

18. Dana is making a batch of cookies and needs to measure the flour. Which measuring tool should she use?

19. Lacy is measuring the length of her marker. Which unit should she use to record her measurement?

20. Which unit would you use to measure the mass of a roll of pennies?

A kilograms **C** millimeters
B grams **D** liters

21. Which tool is best for measuring the length of your little finger?

F centimeter ruler **H** thermometer
G measuring cup **J** balance scale

Name ______________________ **Lesson 17.2**

Metric Length

Estimate the length of the following in millimeters, centimeters, or meters. Then measure and record your actual measurement.

1. Height of the back of your chair

   Estimate: ______________

   Measurement: ______________

2. Length of your longest fingernail

   Estimate: ______________

   Measurement: ______________

3. Length of your shoe

   Estimate: ______________

   Measurement: ______________

4. Diameter of a pencil eraser

   Estimate: ______________

   Measurement: ______________

5. Thickness of your math textbook

   Estimate: ______________

   Measurement: ______________

6. What units did you choose in Exercise 5, and why did you choose them?

   ____

   ____

Write the appropriate metric unit for measuring each.

7. Distance from your house to your school ______________

8. Width of a dictionary ______________

9. Height of the ceiling in your classroom ______________

10. Length of an apple stem ______________

11. Distance from Reno to Minneapolis ______________

12. Width of a key on a computer keyboard ______________

Estimate and measure each.

13. 

   Estimate: ______________

   Measurement: ______________

14. 

   Estimate: ______________

   Measurement: ______________

15. 

   Estimate: ______________

   Measurement: ______________

16. 

   Estimate: ______________

   Measurement: ______________

17. 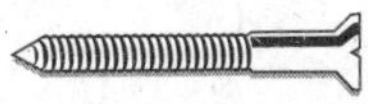

   Estimate: ______________

   Measurement: ______________

Name____

Change SI (Metric) Linear Units

Change the unit.

1. 12 m = ☐ cm
2. 24 mm = ☐ cm
3. 9 km = ☐ m
4. 19 cm = ☐ m
5. 2,500 m = ☐ km
6. 77 mm = ☐ m
7. 59 m = ☐ cm
8. 0.03 m = ☐ cm
9. 3,340 cm = ☐ m
10. 0.001 km = ☐ cm
11. 43,400 cm = ☐ km
12. 19 mm = ☐ cm

Add or Subtract.

13. 110 cm + 1.1 m ________
14. 65 mm + 34.3 cm ________
15. 1.04 m − 6 cm ________
16. 88 cm − 0.56 m ________
17. 12 cm + 19 mm ________
18. 3.09 km − 221 m ________
19. 98 mm − 9.8 cm ________
20. 0.225 km + 22.5 m ________
21. 1.04 m + 74 cm ________
22. 5 km + 1,300 m ________
23. 279 mm − 0.6 cm ________
24. 4,400 cm − 8,000 mm ________

Problem Solving and TAKS Prep

25. Piper combined two 45 cm chains and one 36 cm chain to make one chain, how many meters long is the new chain?

____

26. Howie cut 34.5 cm off a 1-meter board. How much board is left?

____

27. How many millimeters are in 110 centimeters?

A 1,100 mm
B 110 mm
C 11 mm
D 1.1 mm

28. A length of 1.2 km equals which of the following?

F 12 meters
G 120 meters
H 1,200 meters
J 12,000 meters

Name ____________________

Metric Capacity and Mass

Change the unit.

1. 80 L = ☐ kL
2. 900 mg = ☐ g
3. 3 metric cups = ☐ mL
4. 18,000 mL = ☐ L
5. 5 kg = ☐ g
6. 130 ml = ☐ L
7. 336 g = ☐ mg
8. 8.25 L = ☐ mL
9. 1,200 mg = ☐ g

Find the sum or difference.

10. 12 mg + 12 mg = ____________
11. 0.7 kL − 0.6 kL = ____________
12. 20 ml − 0.2 ml = ____________
13. 12 g + 1,100 g = ____________

Find the missing measurement.

14. 1 L + ☐ = 1 kL
15. 140 mg + ☐ = 1.2 g
16. 500 ml − ☐ = 2 metric cups
17. 300 g + ☐ = 0.5 kg
18. 4 g − ☐ = 250 mg
19. 1 L − ☐ = 2 ml

Problem Solving and TAKS Prep

20. Jenna and Annie are making applesauce and need 5 kilograms of apples. How many grams are in 5 kilograms?

____

21. Cal drank 800 milliliters of water at school today and 500 milliliters at home. How many liters did Cal drink?

____

22. Kennedy's dog weighs 34,000 g. How many kilograms does Kennedy's dog weigh?

A 3,400 kg
B 340 kg
C 34 kg
D 3.4 kg

23. Carson's water bottle holds 1.5 liters. What is the capacity of his water bottle in metric cups?

F 3 metric cups
G 4 metric cups
H 5 metric cups
J 6 metric cups

Name____ **Lesson 17.5**

Problem Solving Workshop Skill: Estimate or Actual Measurement

Problem Solving Skill Practice

Tell whether you need an estimate or an actual measurement. Then solve.

1. Gianna is making pendant necklaces for 5 of her friends. She has a spool that has 2.2 m of leather string. If she needs 42 cm of leather string for each necklace, does Gianna have enough?

____

____

2. Dominic is making a birdhouse and needs to cut 3 pieces of trim that are 14, 31, and 44 cm long. Dominic has one 1-meter-long piece of trim. Is it long enough?

____

____

Mixed Applications

USE DATA For 3–5, use the table.

3. Frannie is shopping for beading materials. She wants to make 23 20-cm bracelets with silver wire. Will one spool of silver wire be enough?

Stringing Materials	
Satin cord	$2.89/10 meters
Elastic thread	$2.31/10 meters
Silver wire	$2.50/10 meters
Silk thread	$8.63/10 meters

____

4. Brooklyn wants to make four 45-cm necklaces. If the store will let her buy her stringing material by the meter instead of by the spool, how many meters should Brooklyn ask for?

____

5. Jeff and Mia are making 120 necklaces to sell at their school's craft sale. One-third of the necklaces will be 30 cm long, one-third will be 40 cm long, and the rest will be 50 cm long. How many spools of thread will Jeff and Mia need to buy?

____

Name ____________________

Fahrenheit Temperature

Find the change in temperature.

1. 56°F to 20°F ________

2. 7°F to $^-$17°F ________

3. 88°F to 101°F ________

4. 16°F to 20°F ________

5. $^-$6°F to 20°F ________

6. 100°F to 10°F ________

7. 19°F to 9°F ________

8. 4°F to $^-$20°F ________

9. $^-$2°F to 2°F ________

10. 107°F to 98°F ________

11. 90°F to 9°F ________

12. $^-$29°F to 49°F ________

13. 11°F to 7°F ________

14. 79°F to 90°F ________

15. $^-$32°F to 0°F ________

16. 19°F to 99°F ________

17. 67°F to 76°F ________

18. $^-$8°F to 88°F ________

Problem Solving and TAKS Prep

19. In Paris, the temperature is 82°F, and in Boston, it is 68°F. What is the temperature difference between the two cities?

____

20. If the refrigerator is 38°F and the freezer is −2°F, what is the difference in temperature?

____

21. The temperature at 5 P.M. was 77°F. By midnight, it was 60°F. What was the change in temperature?

A 15°F

B 16°F

C 17°F

D 18°F

22. At 6 A.M., it was −2°F. By noon, it was 18°F. What was the change in temperature?

F 22°F

G 20°F

H 18°F

J 16°F

Name____________________ **Lesson 17.7**

Celsius Temperature

Find the change in temperature.

1. 6°C to 20°C ________

2. 10°C to $^{-}7$°C ________

3. 18°C to 40°C ________

4. 16°C to 0°C ________

5. $^{-}6$°C to 2°C ________

6. 30°C to 10°C ________

7. 1°C to 9°C ________

8. 4°C to $^{-}35$°C ________

9. $^{-}2$°C to 2°C ________

10. 17°C to 8°C ________

11. 0°C to $^{-}9$°C ________

12. $^{-}29$°C to $^{-}4$°C ________

13. 11°C to 7°C ________

14. 9°C to 15°C ________

15. $^{-}40$°C to 40°C ________

16. 12°C to $^{-}9$°C ________

17. 7°C to $^{-}7$°C ________

18. $^{-}8$°C to 13°C ________

Problem Solving and TAKS Prep

19. In Atlanta, the temperature is 27°C, and in Anchorage, it is −5°C. What is the temperature difference between the two cities?

____

20. The temperature inside the refrigerator is 4.4°C. How many degrees Celsius above freezing is the inside of the refrigerator?

____

21. It is 2°C in Denver and 20°C in Dallas. How many degrees Celsius separate the two cities?

A 22°C

B 18°C

C 14°C

D 10°C

22. At midnight, it was −16°C, and at noon, it was 6°C. What was the change in temperature in degrees Celsius?

F 22°C

G 18°C

H 14°C

J 10°C

Name______________________________

Estimate Perimeter

1. Trace around the outline of a rectangular object in the space below. Then use string and a ruler to estimate the perimeter in centimeters.

____

2. Using string and a ruler, estimate the perimeter of your math book.

____

Estimate the perimeter of each polygon in centimeters.

3.

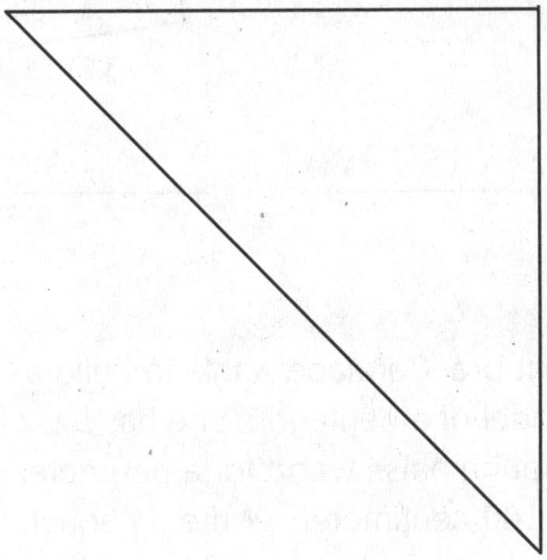

____

4.

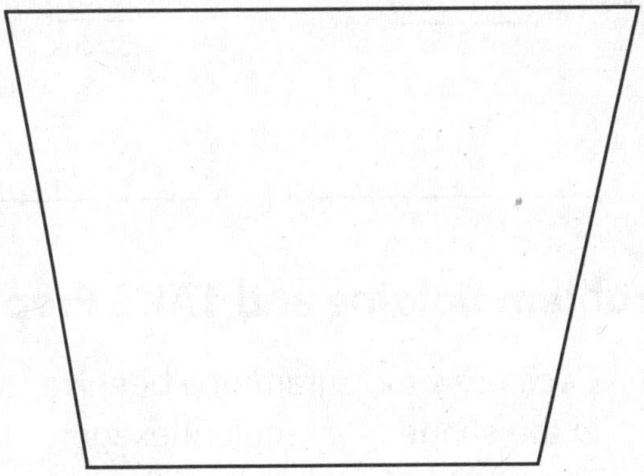

____

5.

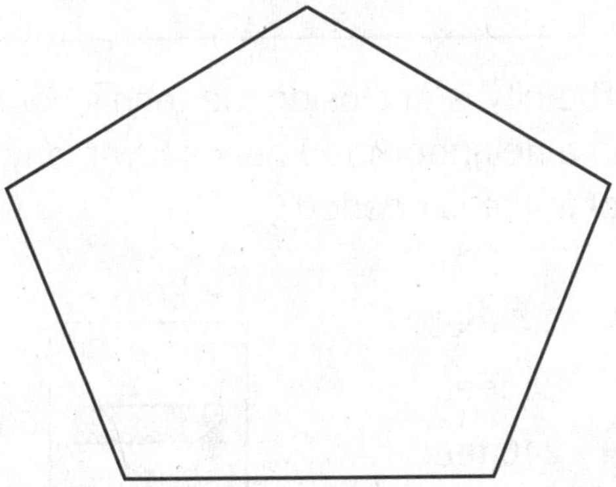

____

6.

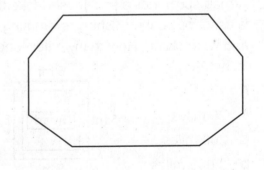

____

Name____

Lesson 18.2

Find Perimeter

Find the perimeter of each polygon.

1.
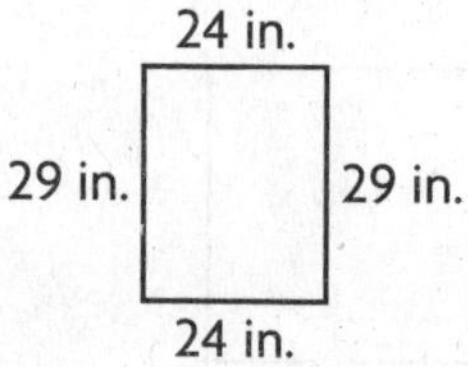

2.
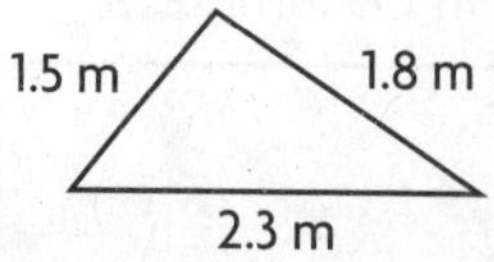

3.
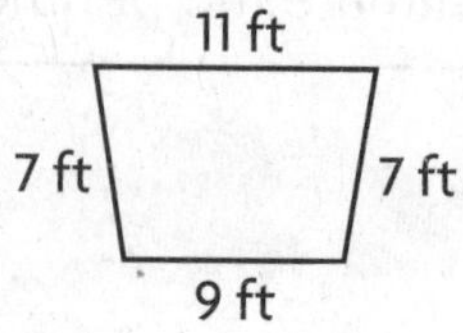

4.
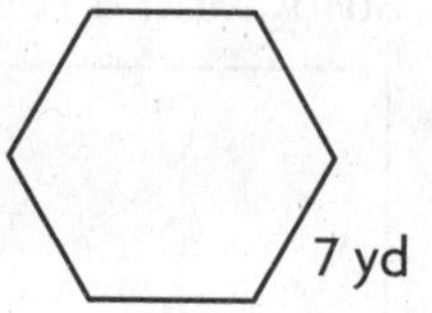

5.
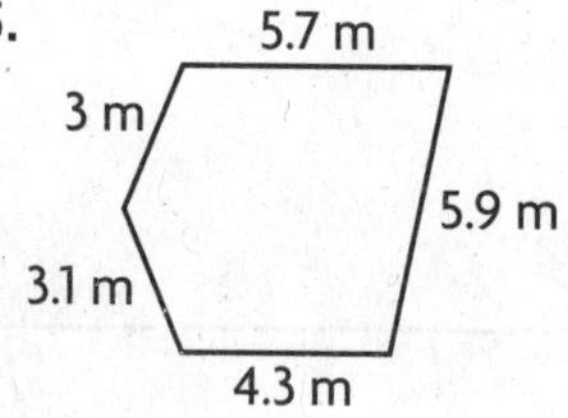

6.
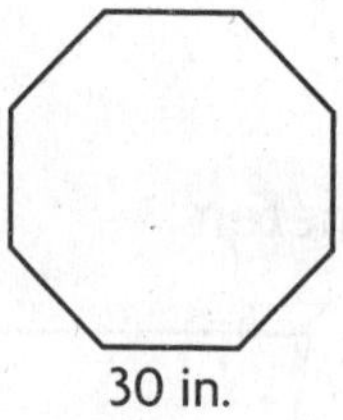

7.
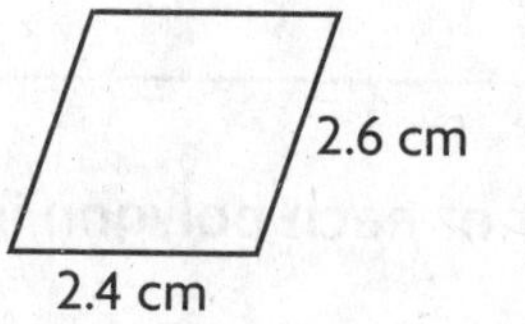

8.
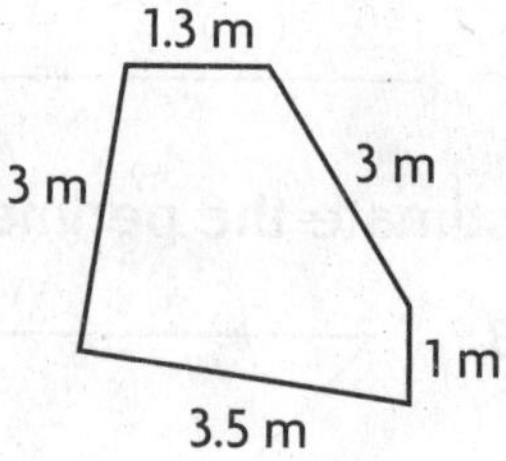

Problem Solving and TAKS Prep

9. Cecil drew a diagram of a beehive in the shape of a regular hexagon. The length of each side of the hexagon was 4.5 inches. What was the perimeter of Cecil's diagram?

10. **Algebra** Candace wants to build a model of a pentagon. She has enough balsa wood for a perimeter of 100 centimeters. Write an equation Candace could use to find the length of each side. Then solve the equation.

11. A road surrounds a small resort lake that is used for boating, fishing, swimming, and waterskiing. How many miles long is the road?

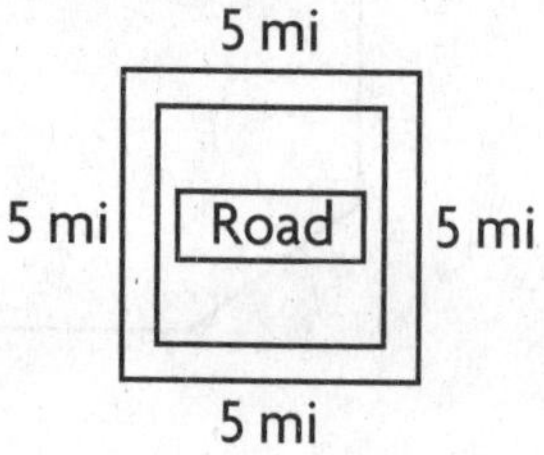

A 2 miles
B 20 miles
C 200 miles
D 2,000 miles

12. The city is enclosing the tennis court in a neighborhood park. How many feet of fence is needed?

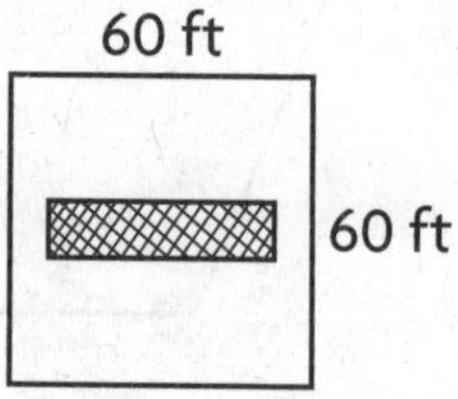

F 2.4 feet
G 24 feet
H 240 feet
J 2,400 feet

Name____

Algebra: Perimeter and Formulas

Find the perimeter of each polygon by using a formula.

1.
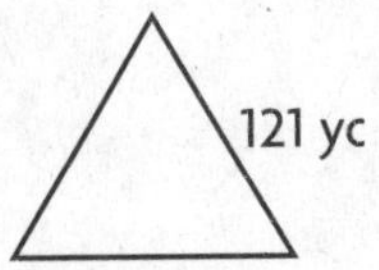

____

2.
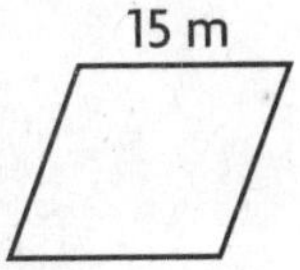

____

3.
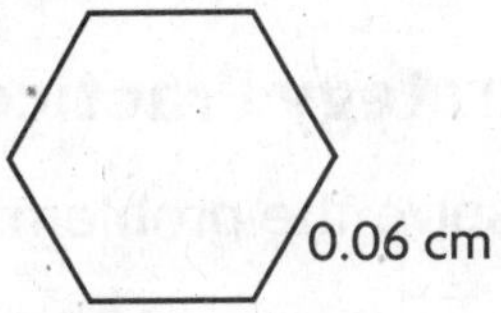

____

4.
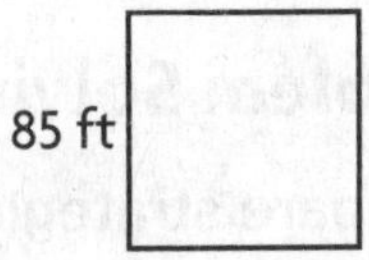

____

5.
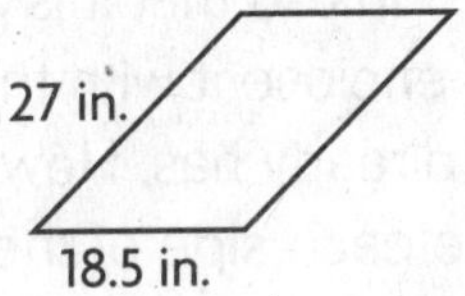

____

6.
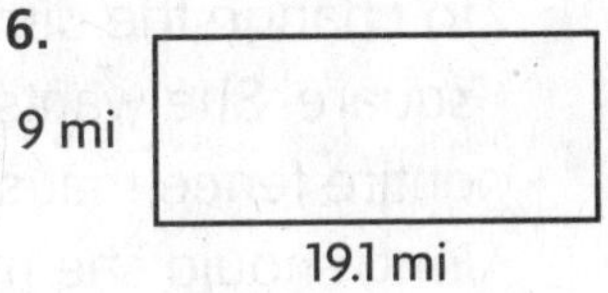

____

7.
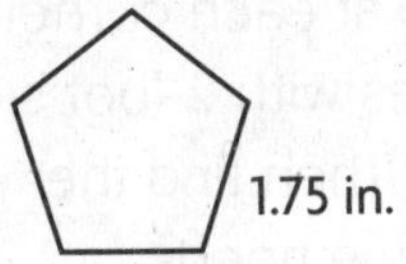

____

8.
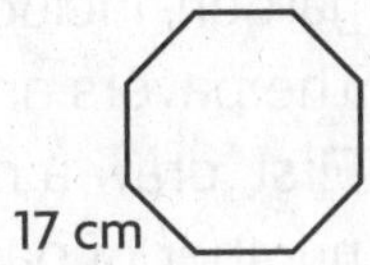

____

9.
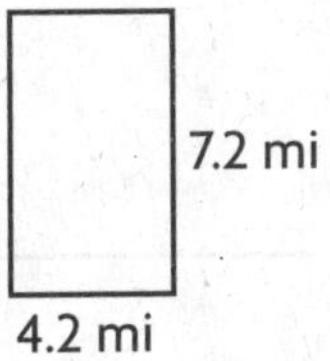

____

10.
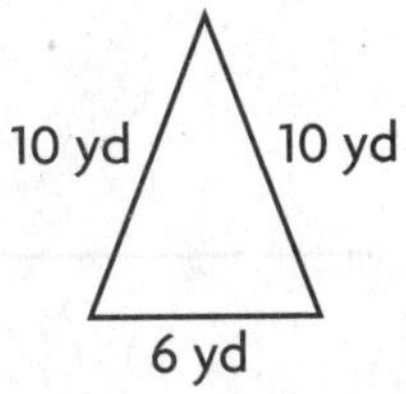

____

11.
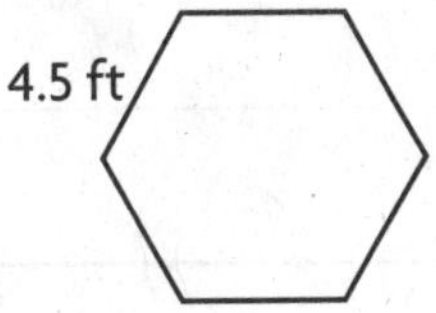

____

12.
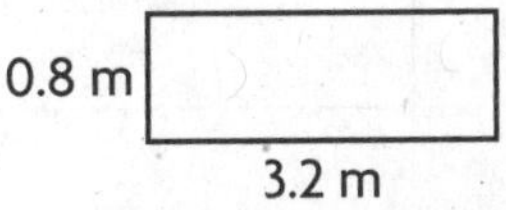

____

Problem Solving and TAKS Prep

13. **Algebra** The perimeter of a regular hexagon is 42 yards. What is the length of each side?

____

14. The side chambers of the Lincoln Memorial are each 38 feet wide and 63 feet long. What is the perimeter of a side chamber?

____

15. For which polygon could you use the formula $P = 2l + 2w$ to find its perimeter?

A triangle
B rectangle
C trapezoid
D pentagon

16. For which regular polygon could you use the formula $P = 5x$ to find its perimeter?

F triangle
G square
H pentagon
J hexagon

Name____

Problem Solving Workshop Strategy: Compare Strategies

Problem Solving Strategy Practice

Compare strategies to solve the problem.

1. Shane has a rectangular garden that is 4 feet wide and 12 feet long. He wants to buy stone pavers to go around the garden, including one at each corner. The pavers are squares with 2-foot sides. First, draw a diagram. Then find the number of pavers Shane needs.

____

2. Lucy has a vegetable garden that is 12 feet wide and 16 feet long. She has enclosed it with a fence. Now she wants to change the dimensions so that it is a square. She wants to enclose it with the entire fence that she already has. How long should she make each side of the square garden?

____

Mixed Applications

USE DATA For 3–4, use the diagram of The Alamo.

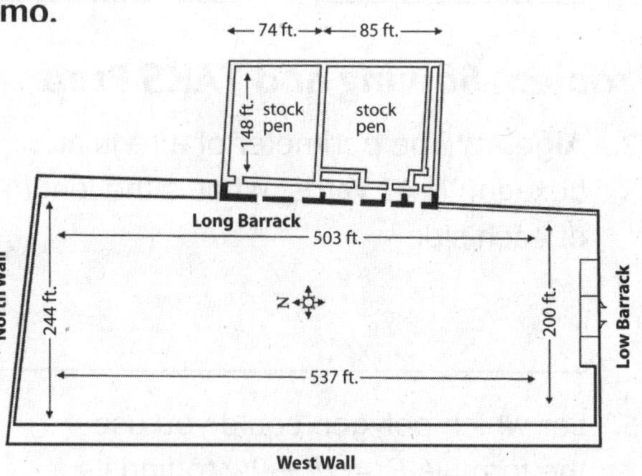

3. What is the perimeter of The Alamo, including the outside wall of the two stock pens?

____

4. Which stock pen has the greater perimeter?

____

5. In June, Layne lifted weights on every date evenly divisible by 3 and jogged on every date evenly divisible by 4. On which dates did she do both?

____

6. Jake's car gets about 25 miles per gallon. His tank holds 12 gallons, but he refuels when the tank is a quarter full. He just filled his tank. About how many miles can he travel before refueling?

____

Name________________________________

Estimate Area

Estimate the area of the shaded figure. Each square on the grid is 1 cm^2.

1.

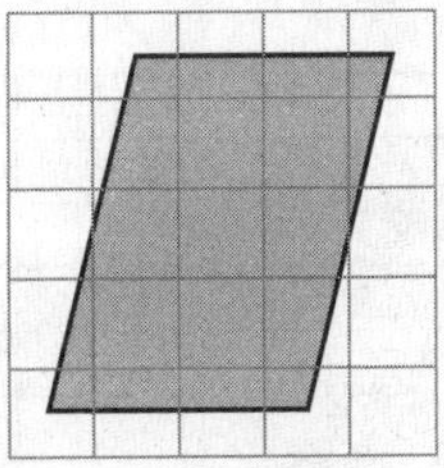

____

2.

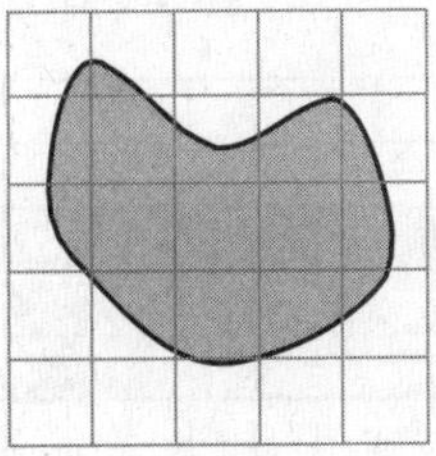

____

3.

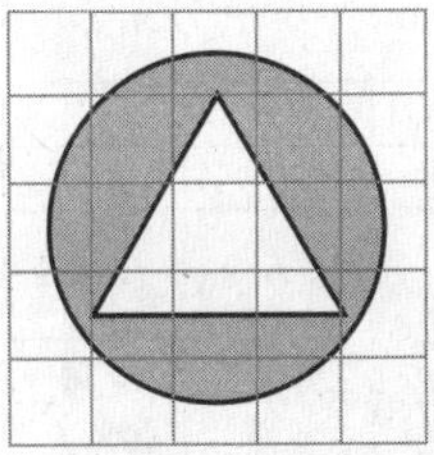

____

Problem Solving and TAKS Prep

USE DATA For 4–5, use the train puzzle on the right.

Train Puzzle (each square is 1 inch)

4. The jigsaw puzzle of a train at the right has 100 pieces. Estimate the area of the puzzle.

____

5. Estimate the area of the train in the jigsaw puzzle at the right.

____

6. Which is a reasonable estimate for the area of the figure?

A 15 $in.^2$

B 9 $in.^2$

C 4 $in.^2$

D 2 $in.^2$

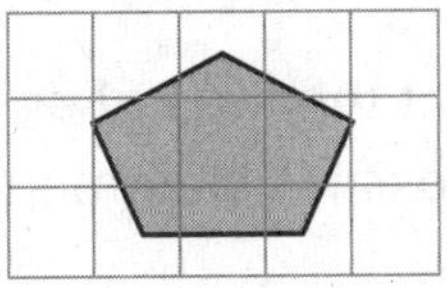

□ = 1 $in.^2$

7. Which is a reasonable estimate for the area of the banner?

F 4 cm^2

G 8 cm^2

H 12 cm^2

J 15 cm^2

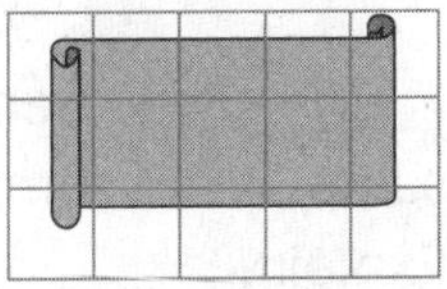

□ = 1 $cm.^2$

Name

Explore Areas of Squares and Rectangles

Count the number of shaded squares to find the area of each figure. Each square represents 1 square inch.

1. 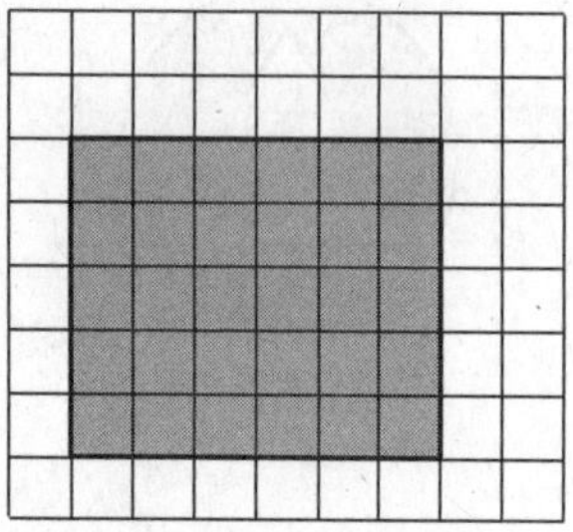

2. 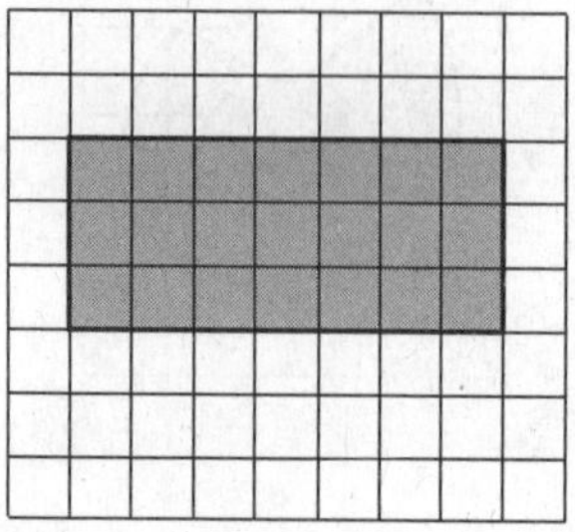

3. 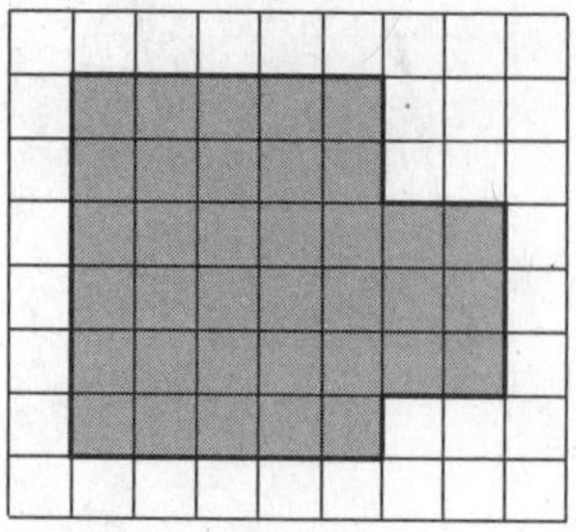

Draw each figure on the grid below. Find the area.

4.  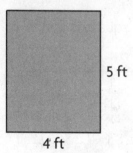

5. 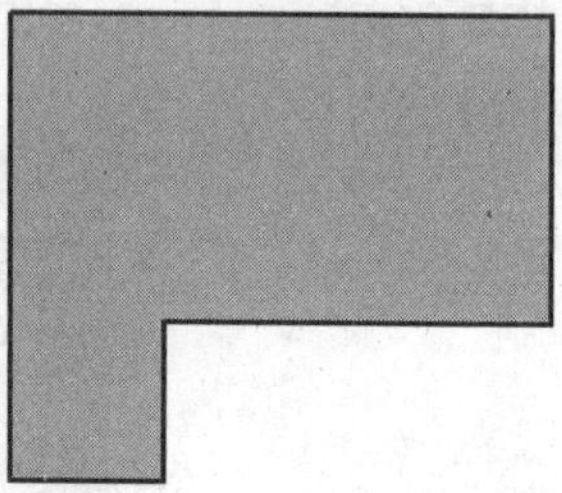

6. 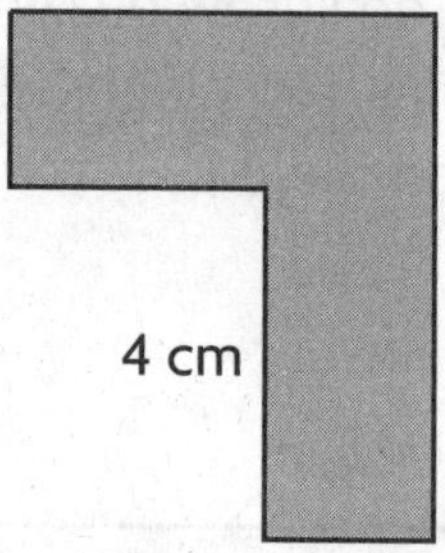

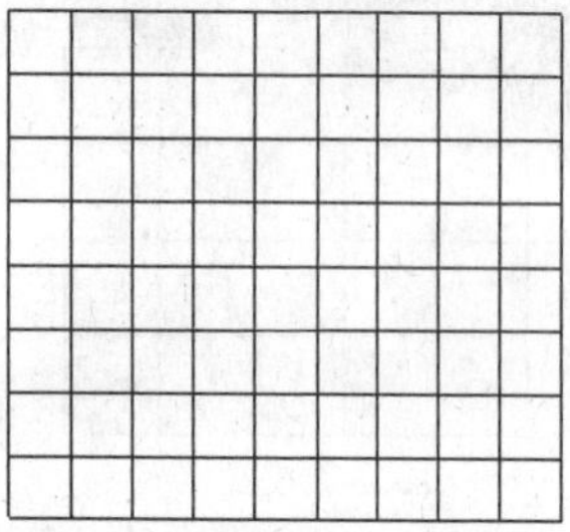

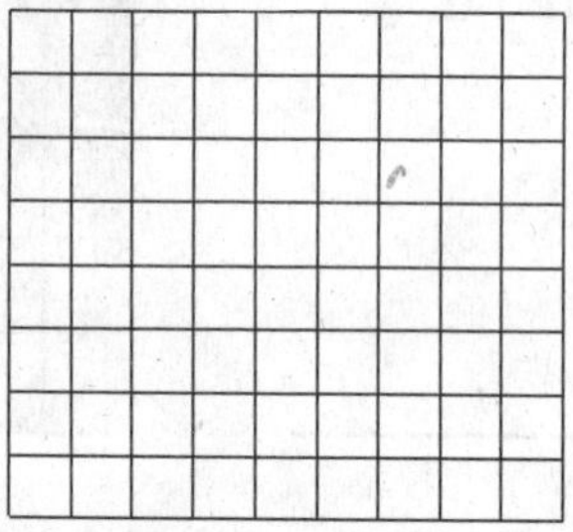

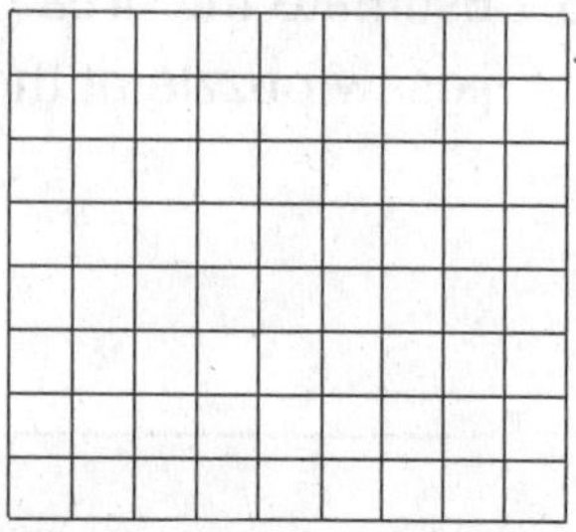

7. WRITE Math **Sense or Nonsense** Josiah says that the area of an irregular shape is the same as the sum of the areas of its parts. Does Josiah's statement make sense? **Explain.**

Name ______________________________

Algebra: Area of Rectangles

Find the area of each figure.

1.

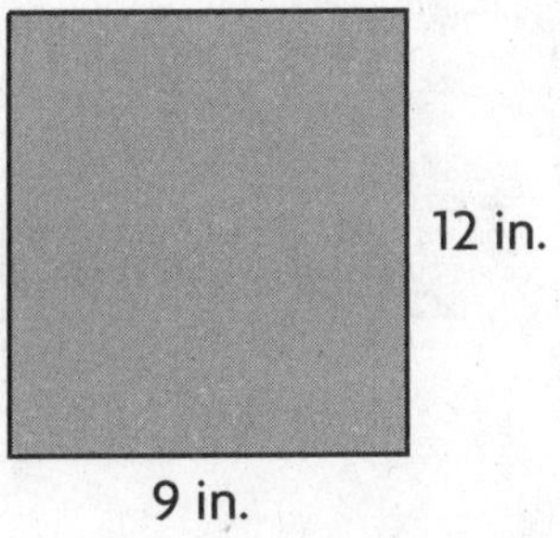

____

2.

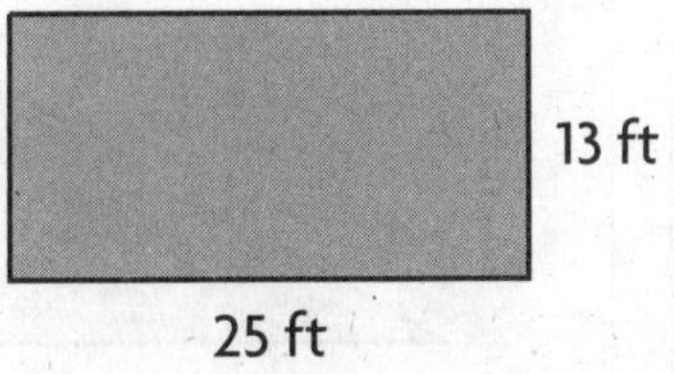

____

3.

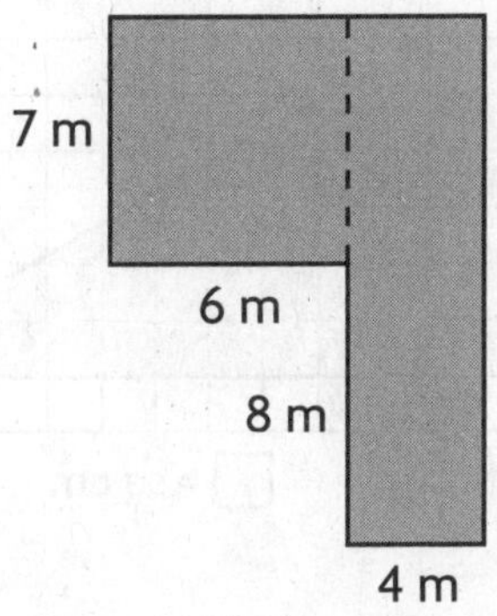

____

Problem Solving and TAKS Prep

USE DATA For 4–5, use the table.

Room	Length	Width
Living room	25 ft	15 ft
Dining room	18 ft	14 ft
Kitchen	12 ft	16 ft

4. Stephanie is buying carpet for the living room. How many square feet of carpet does Stephanie need in order to cover the entire floor?

____

5. Jonathon plans to replace all 252 square feet of flooring in one of the three rooms. Which room does Jonathon plan to replace the floor?

____

6. How many 1 square foot tiles would you need to cover a 16 ft × 18 ft kitchen?

A 68 tiles

B 144 tiles

C 288 tiles

D 300 tiles

7. How many square feet of carpet are needed to carpet a living room whose length is 27 ft and width is 20 ft?

F 94 ft^2

G 540 ft^2

H 500 ft^2

J 640 ft^2

Name______________________________

Algebra: Area of Triangles

Find the area of each triangle.

1.

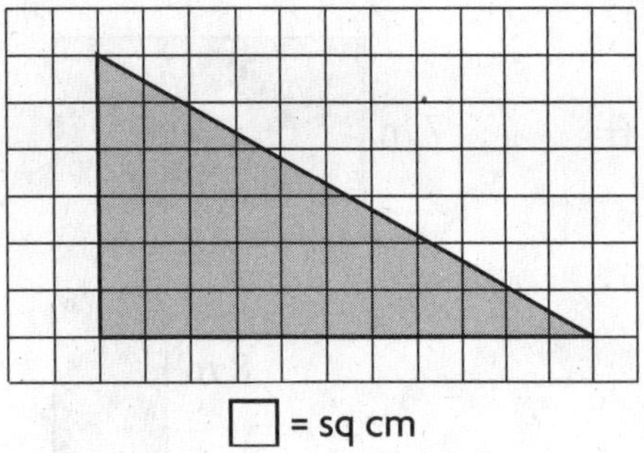

____

2.

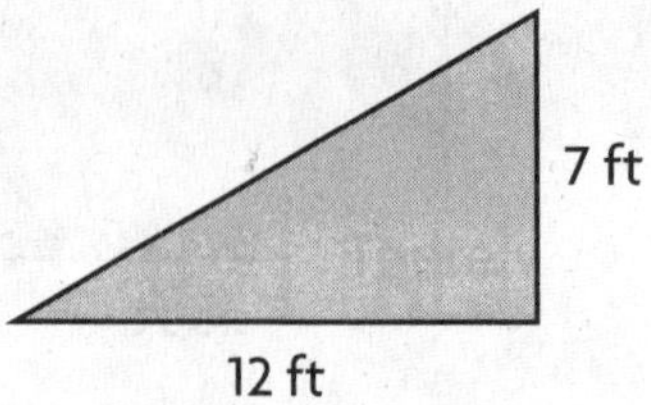

____

3.

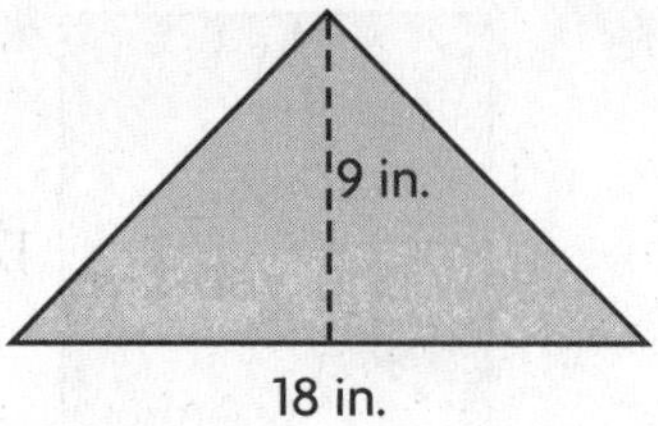

____

Problem Solving and TAKS Prep

USE DATA For 4–5, use the pattern.

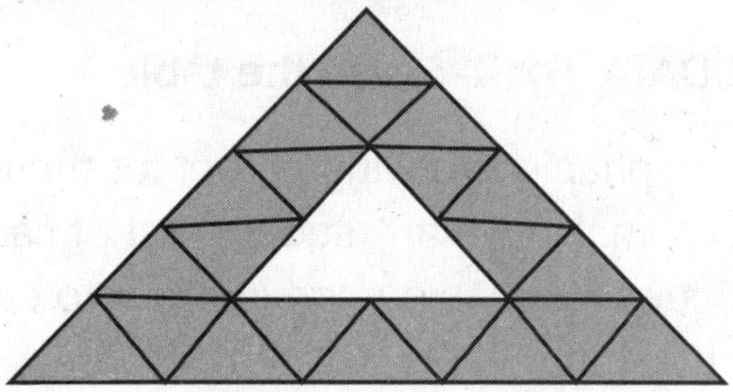

4. Kate bought blue tiles to fill the middle of the pattern. How many blue tiles did she buy?

____

5. **Reasoning:** The tiles in the pattern are right isosceles triangles. The two shorter sides of each triangle are each 2 cm long. Estimate the area of the shaded part of the pattern.

____

6. Mark wants to fence a rectangular area of 240 square feet for his dog. If the fence is 24 feet long, how wide will the fence be?

____

7. What is the area of the triangular flag shown below?

A 15 ft^2

B 28 ft^2

C 30 ft^2

D 56 ft^2

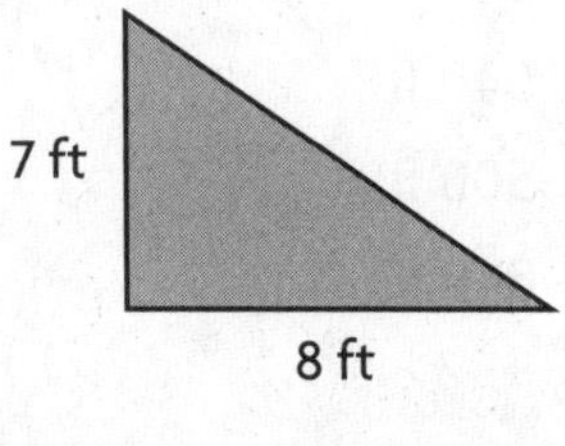

8. A triangular figure has a height of 12 cm and a base of 5 cm. What is the area of the triangular figure?

F 17 cm^2

G 22 cm^2

H 30 cm^2

J 60 cm^2

Name ____________________

Lesson 19.5

Algebra: Area of Parallelograms

Find the area of each parallelogram.

1.

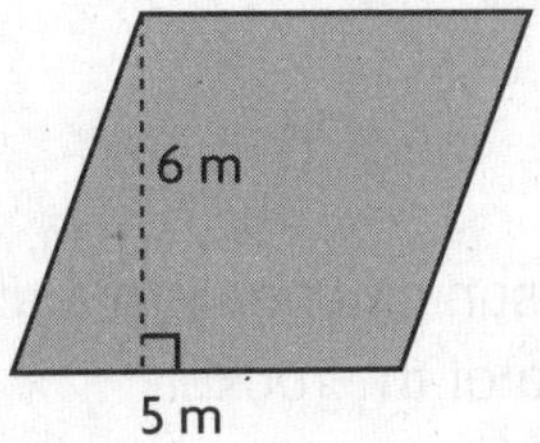

____

2.

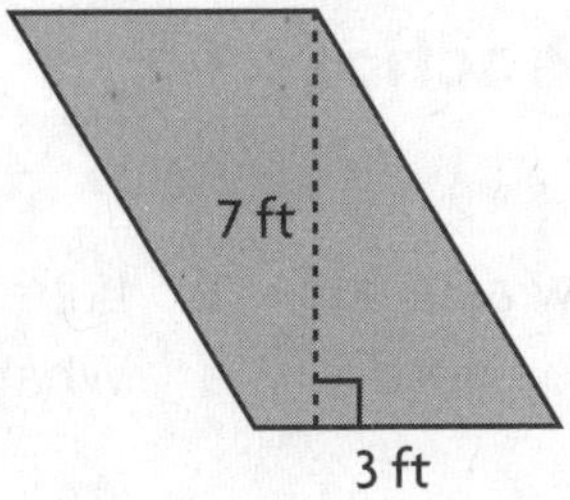

____

3.

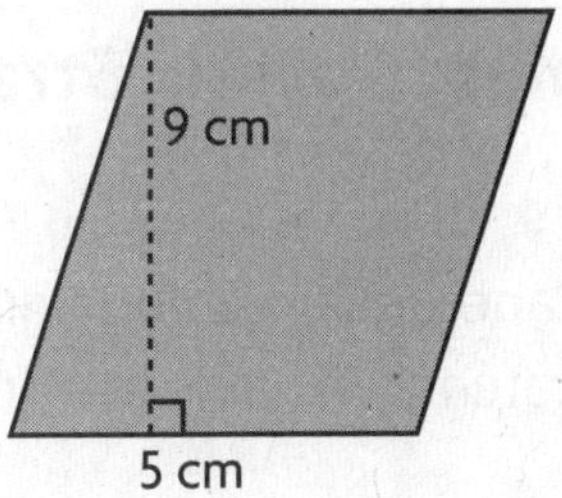

____

4.

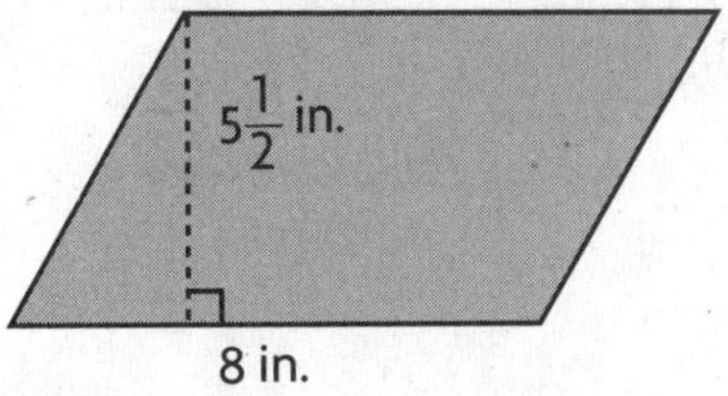

____

5.

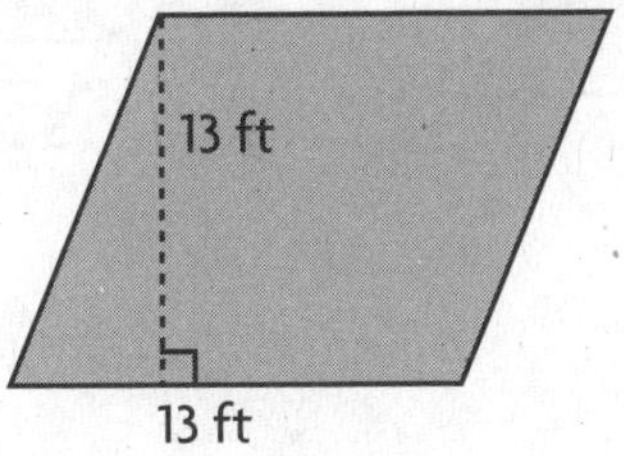

____

6.

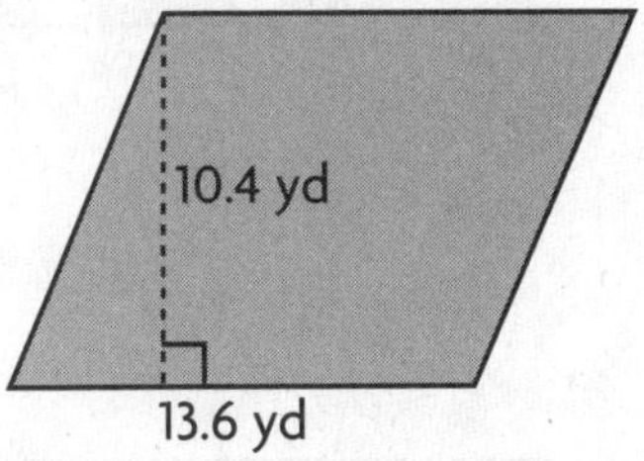

____

Problem Solving and TAKS Prep

7. A yard is shaped like a parallelogram with a base of 27 m and a height of 30 m. What is the area of the yard?

____

8. A parallelogram has a length of 15 cm and a height of 20 cm. It is divided into two congruent triangles. What is the area of each triangle?

____

9. What is the area of the parallelogram?

A 300 yd^2

B 70 yd^2

C 294 yd^2

D 147 yd^2

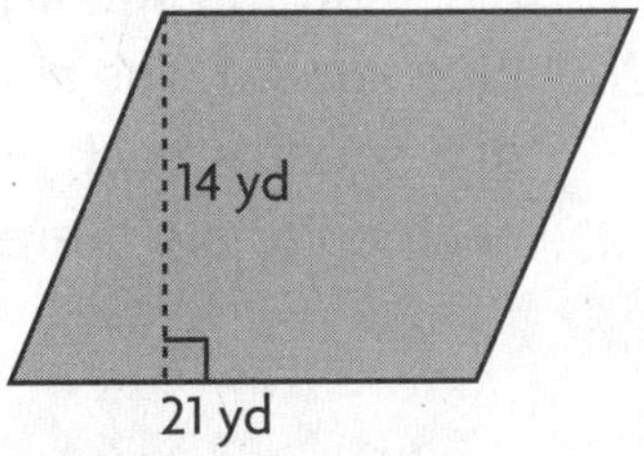

10. A playground is divided into two equal parallelograms. What is the area of the entire playground? Show your work.

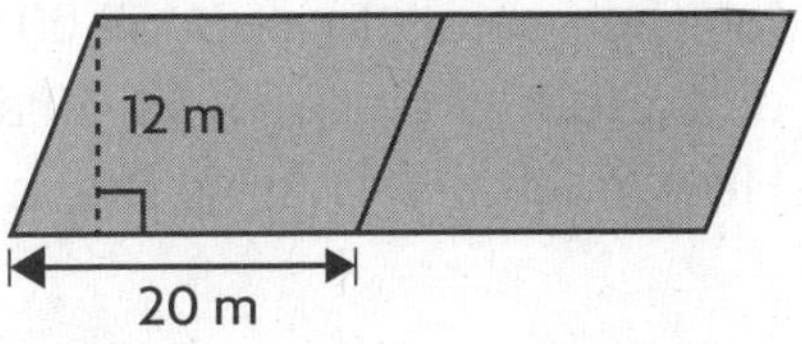

____

Name____

Problem Solving Workshop Strategy: Solve a Simpler Problem

Problem Solving Strategy Practice

Solve.

1. Jane designed the figure below as a sun catcher. What is the area of the figure?

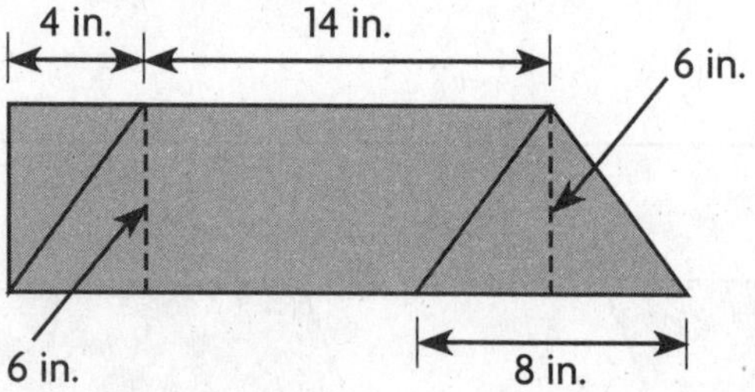

____

2. Luke made his sun catcher into a rocket. What is the area of the rocket?

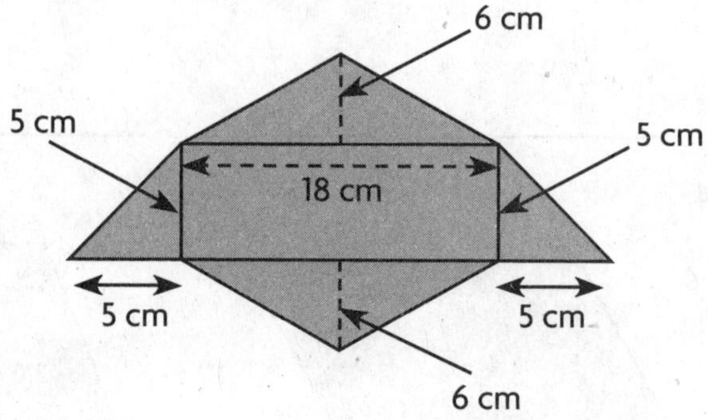

____

Mixed Strategy Practice

USE DATA For 3–4, use the diagram.

3. Chris designed his sun catcher to the right into an airplane. What is the area of Chris' airplane?

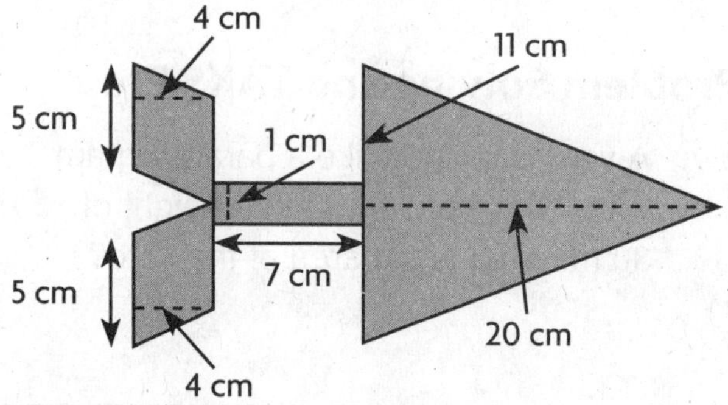

____

4. Chris bought the materials for the sun catcher. He paid $1.50 each for each rectangle, $2.25 for each triangle, $1.75 for each parallelogram, $3.00 for stain and 3 feet of chain for $4.50 a foot. How much did Chris spend in all?

____

5. Joy made a sun catcher with alternating blue and red squares. She began with a blue square. The sun catcher has 9 rows of 5 squares each. How many squares of each color are there?

____

Name____ **Lesson 20.1**

Measure Volume

Find the volume of each rectangular prism.
You may want to use the net to make the prism.

1.

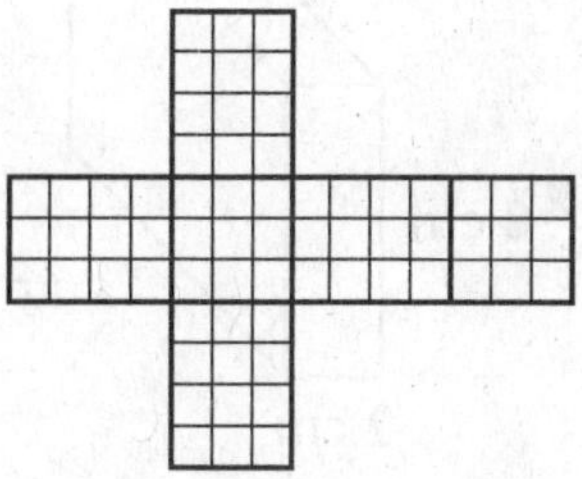

____

2.

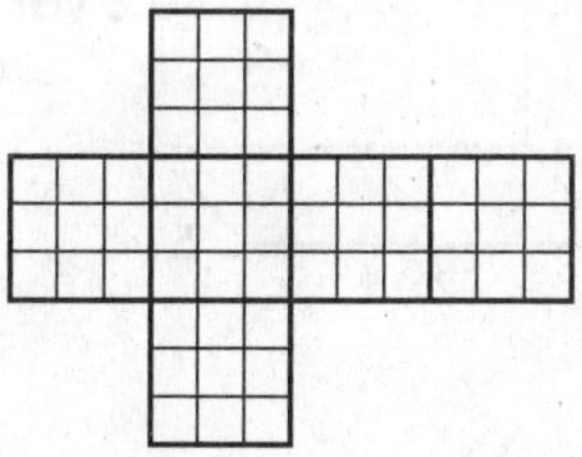

____

3.

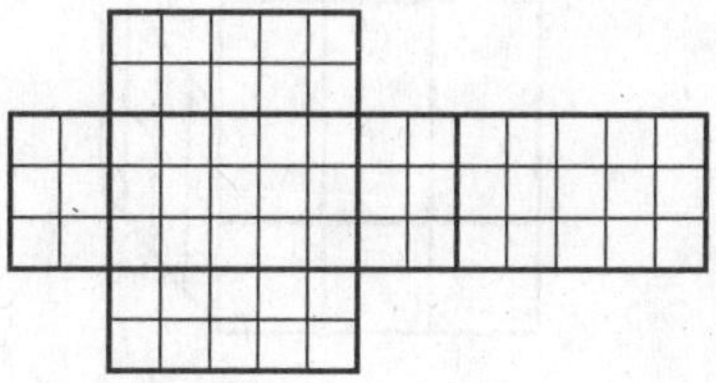

____

Find the volume of each rectangular prism in cubic centimeters.

4.

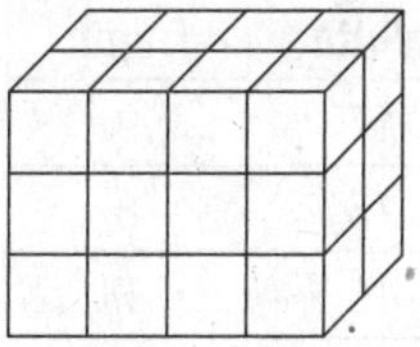

____

5.

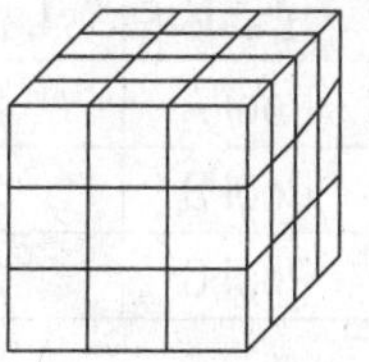

____

6.

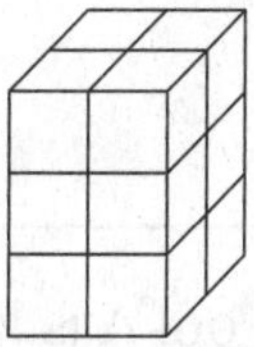

____

7.

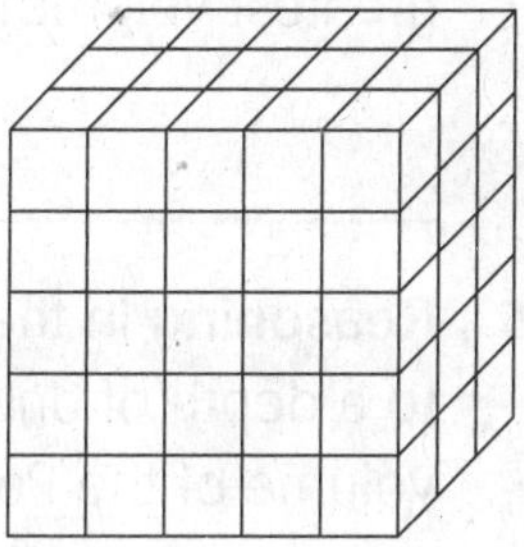

____

8. WRITE Math Explain how to find the volume of a rectangular prism with the dimensions of 7 in. by 5 in. by 4 in.

____

____

____

____

Name________________________________ Lesson 20.2

Algebra: Find Volume

Find the volume of each rectangular prism.

1.

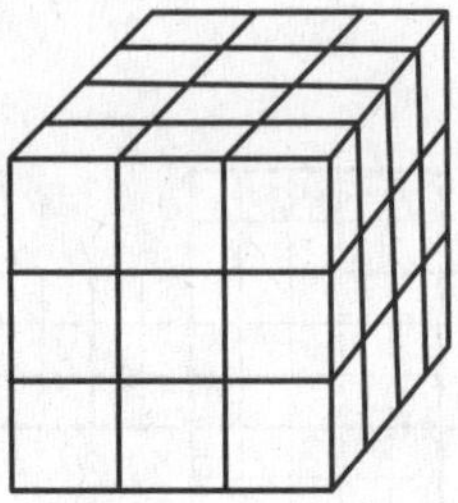

2.

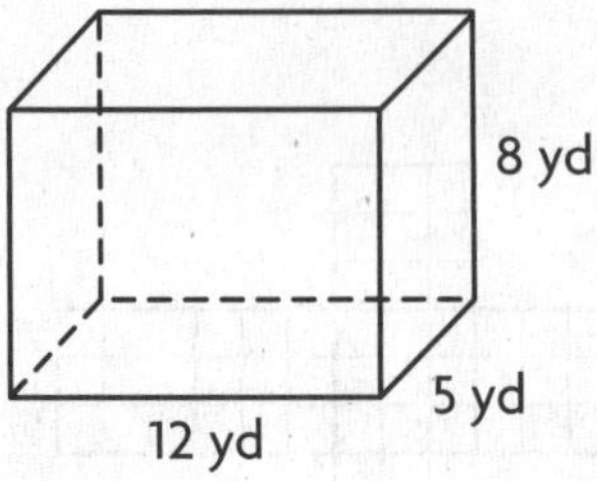

3.

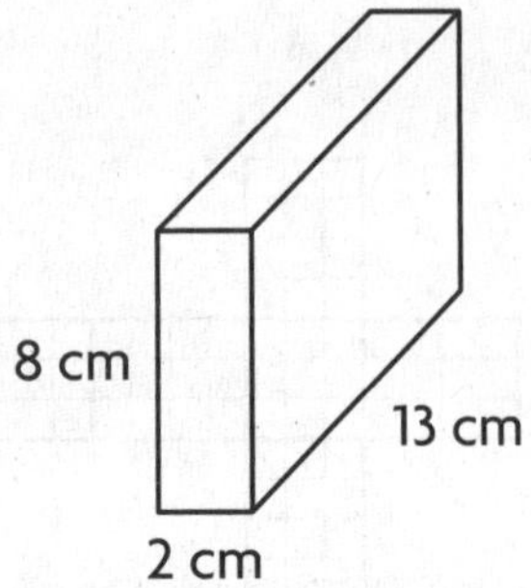

______________ ______________ ______________

Problem Solving and TAKS Prep

For 4–5, use the table.

Swimming Pool Dimensions (in feet)			
Pool	**Length**	**Width**	**Depth**
Pool A	20	17	9
Pool B	25	15	8
Pool C	30	15	7

4. Which of the three pools has the greatest volume?

____

5. **Reasoning** In the winter, Pool A is filled to a depth of only 2 feet. What is the volume of the Pool A?

____

6. A triangle has a base of 8 and a height of 6. What is the area of the triangle?

____

7. The capacity of a garbage bag is the same as a 15 in. by 10 in. by 9 in. box. What is the garbage bags total capacity?

A 150 cu in.

B 135 cu in.

C 1,500 cu in.

D 1,350 cu in.

Name____ **Lesson 20.3**

Relate Perimeter, Area, and Volume

Choose the unit you would use to measure each.

1. tile needed to cover a floor

sq in. or sq ft

2. a door frame

in. or ft

3. the amount of water in a lake

cu cm or cu m

4. wall paper needed to cover a wall

sq ft or sq yd

Write the units you would use for measuring each.

5. area of top of this prism

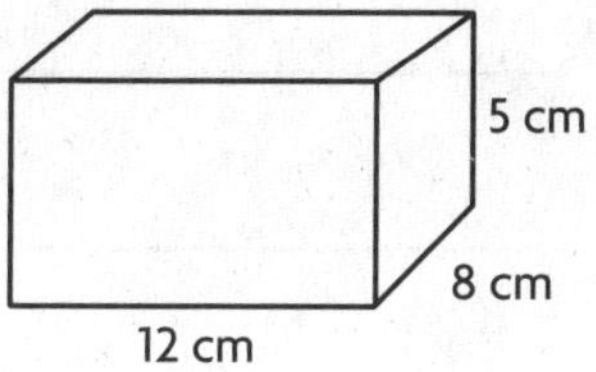

____

6. perimeter of this triangle

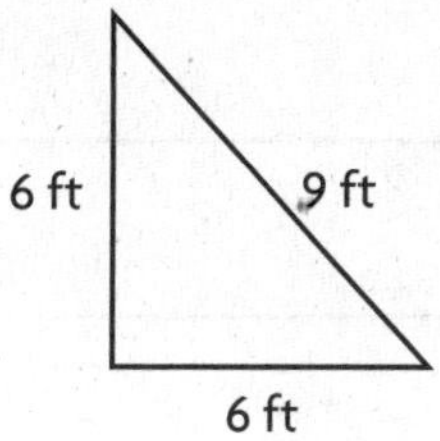

____

7. volume of this prism

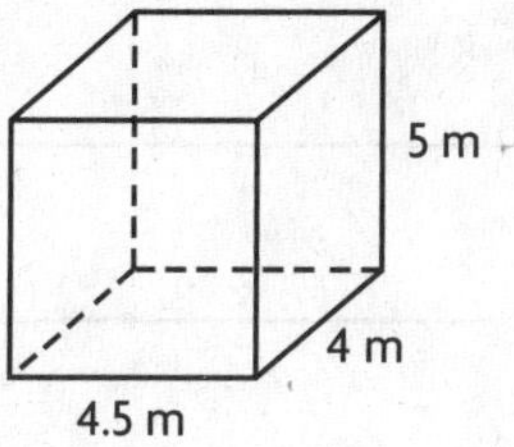

____

Problem Solving and TAKS Prep

USE DATA For 8–9, use the picture of the aquarium.

8. What is the aquarium's volume?

____

9. What is the area of the water's surface that is exposed to the air?

____

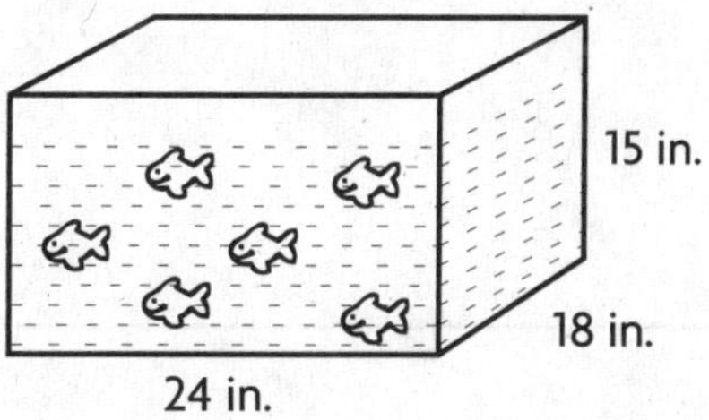

10. Joe wraps the top of a 9 in. by 6 in. by 4 in. box. What units should he use to find how much wrapping paper he needs?

A inches

B square feet

C square inches

D cubic inches

11. Mary wants to fill a 6 in. by 8 in. by 4 in. box with 1 in. blocks. What units should she use to find the number of blocks she needs?

F inches

G square feet

H square inches

J cubic inches

Name________________________

Choose Formulas and Units

Find the perimeter.

1. trim for this flag.

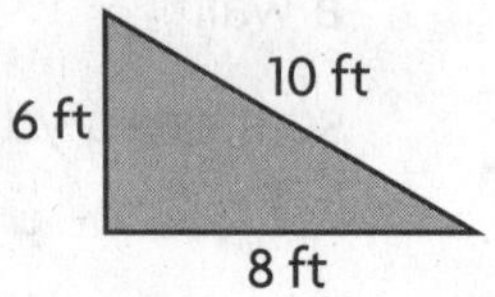

2. crushed stone to cover this path.

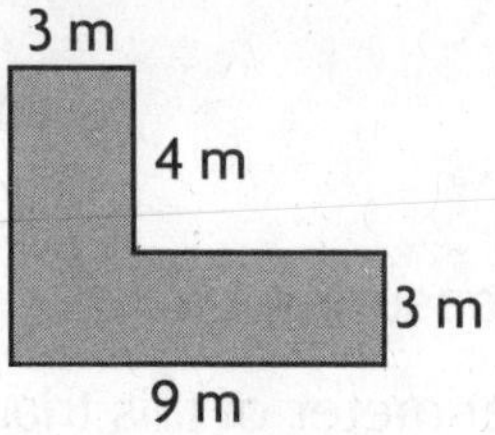

3. *unit cubes in this prism*

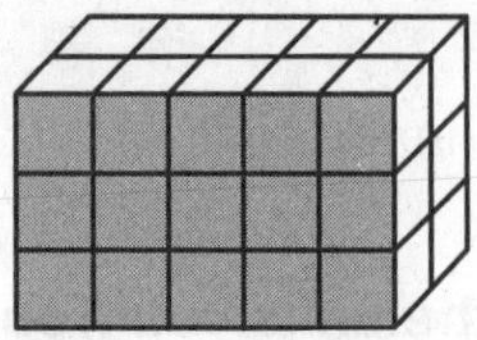

______________ ______________ ______________

______________ ______________ ______________

Problem Solving and TAKS Prep

4. Nathan wants to know how many 1-yard concrete squares are needed to cover a 70-yard by 40-yard picnic area. Choose a formula and solve the problem

____

5. ALGEBRA The area of a painting is 920 square in. Its length is 40 in. Write and solve an equation to find the width of the painting.

____

6. Ben is painting a wall with no windows or doors. What formula will he use to find how much paint he needs?

A $V = l \times w \times h$

B $A = l \times w$

C $A = 6(l \times w)$

D $P = 2l \times 2w$

7. Kelly is building a rectangular sandbox. What formula will she use to find the amount of sand needed to fill the sandbox to the top?

F $V = l \times w \times h$

G $A = l \times w$

H $A = 6(l \times w)$

J $P = 2l \times 2w$

Name____ **Lesson 20.5**

Problem Solving Workshop Skill: Use a Formula

Problem Solving Skill Practice

Tell which formula you would use to solve. Then solve the problem.

1. A bag of lawn fertilizer will cover 135 square feet. Will 5 bags be enough to cover a yard 24 feet long and 23 feet wide? Explain.

____

____

____

2. One tablet of chlorine is used for chlorinating 225 cubic meters of water. Will 10 tablets be enough to chlorinate a 50 m long, 25 m wide and 2 m deep Olympic size pool? Explain

____

____

____

Mixed Applications

USE DATA For 3–4, use the picture.
Julie is making gift boxes for her three younger brothers.

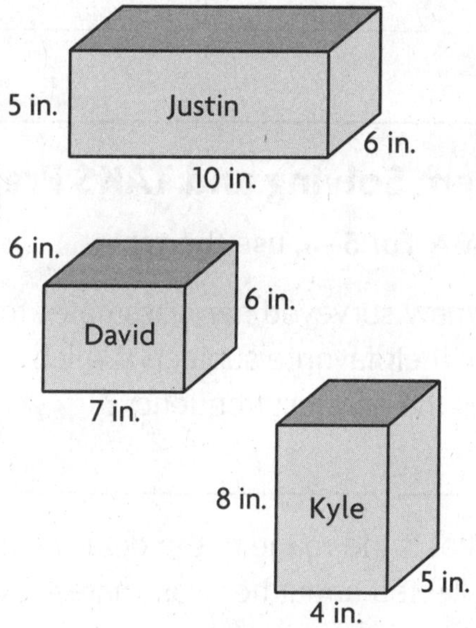

3. Julie is painting stripes around the border along the top of Justin's box. What formula will Julie use? What will the length be?

____

4. Whose gift box has the greatest amount of space inside? Whose has the least?

____

____

____

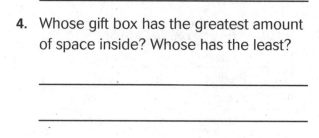

5. Karen is having 156 square feet of wall-to-wall carpeting installed in his living room. His living room is 13 feet wide. What is its length?

____

6. The storage compartment in a moving truck is 5 yards wide, 7 yards long and has a volume of 210 square yards. How high is the storage compartment?

____

Name____

Collect and Organize Data

A movie maker surveyed children 9–13. Tell whether each sample represents the population. If it does not, explain.

1. a random sample of 400 boys, ages 9–13

____

____

2. a random sample of 400 children, ages 9–13

____

____

3. a random sample of 400 teachers

____

____

Make a line plot. Find the range.

4.

Volunteer Hours Survey	
Number of Hours	**Frequency**
2	4
4	10
5	6
7	2

____

Problem Solving and TAKS Prep

USE DATA For 5–6, use the table.

5. Tammy surveyed her classmates to find out their favorite subjects. Which subject has the greatest frequency?

____

6. What is the range of the data Tammy collected about her classmates' favorite subjects?

____

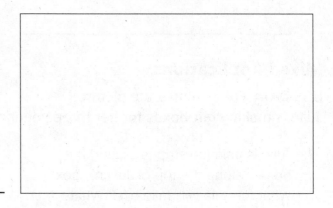

Favorite Subjects	
Spelling	~~////~~
Reading	~~////~~ /
Science	//// ///
Math	~~////~~
Social Studies	//// /

7. What is the range for the following set of data: 14, 9, 11, 21, 7?

A 11

B 12

C 13

D 14

8. Which set of data has a range of 15?

F 4, 9, 2, 15, 18

G 9, 5, 20, 3, 25

H 8, 2, 15, 13, 17

J 5, 20, 7, 14, 21

Name____ **Lesson 21.2**

Median, Mode and Range

For 1–4, use the table. Model the data with connecting cubes.

1. What is the order of the data arranged from least to greatest?

   ____

2. What is the median?

   ____

3. What is the mode?

   ____

4. What is the range?

   ____

Class Size	
Class	**Number of Students**
Math	28
Spelling	25
Phys Ed	22
Science	28
History	27

USE DATA For 5–8, use the table. Model the data with connecting cubes.

5. What is the order of the data arranged from least to greatest?

   ____

6. What is the median?

   ____

7. What is the mode?

   ____

8. What is the range?

   ____

After School Clubs	
Club	**Number of Participants**
Chess	5
Drama	9
French	5
Math	6
Volunteer	7

USE DATA For 9–11, use the line plot.

9. What is the median?

   ____

10. What is the mode?

    ____

11. What is the range?

    ____

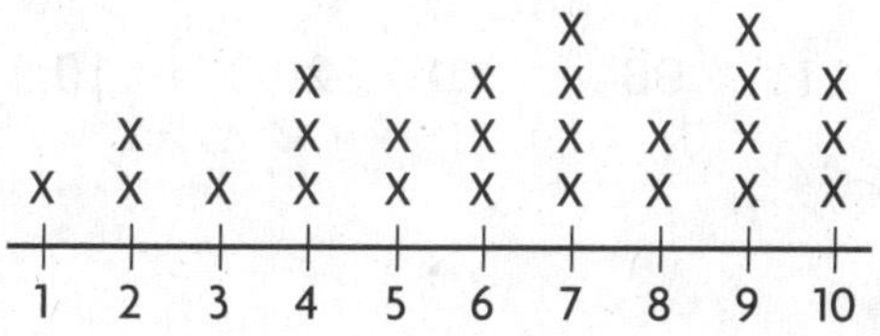

Name________________________________ Lesson 21.3

Compare Data

Compare the data sets. Tell how the data sets compare.

1.

A: Number of stamps collected						
13	25	19	32	66	22	19

B: Number of stamps collected						
6	13	21	20	15	13	24

____

____

____

2.

A: Books Students Read					
14	17	8	15	13	17

B: Books Students Read					
17	10	19	17	18	15

____

____

____

Problem Solving and TAKS Prep

3. Reasoning Hannah and Tyler count the number of times the word *what* occurs. Hananah's data has mean of 2.7 times. What could Tyler's mean be if his results are similar?

____

4. Two data sets have different ranges and medians. Is the data in the data sets similar or different? Explain your reasoning.

____

____

5. Which shows how the median for each set of data compares?
Baseball Cards Saved: 111, 101, 149
Football Cards Saved: 124, 87, 98, 132

A $111 = 111$ **C** $48 > 45$

B $111 > 98$ **D** $120.3 > 110.3$

6. Which shows how the mean for each set of data compares?
Group A Pages Read: 47, 33, 52, 36
Group B Pages Read: 42, 39, 47, 28

F $52 > 47$ **H** $34.5 < 40.5$

G $19 - 19$ **J** $42 > 39$

Analyze Graphs

USE DATA For 1–3, use the double-bar graph.

1. Which class had the least number of right-handed students?

____

2. Which two classes have the same number of students?

____

3. What is the total number of left-handed students in all four classes?

____

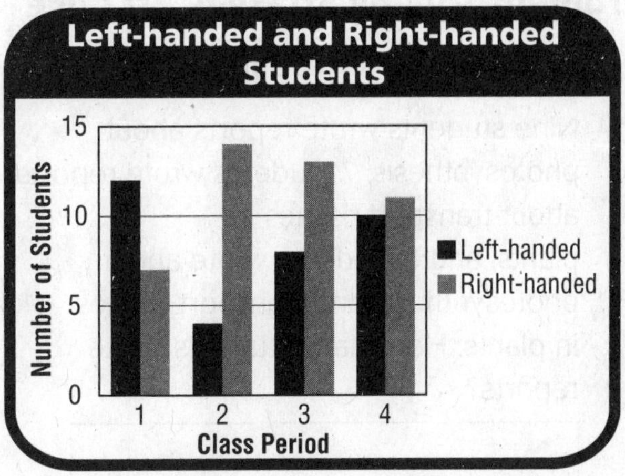

USE DATA For 4–5, use the circle graph.

4. What part of all birds seen are the cardinals?

____

5. Which birds were $\frac{1}{3}$ of the number of birds seen?

____

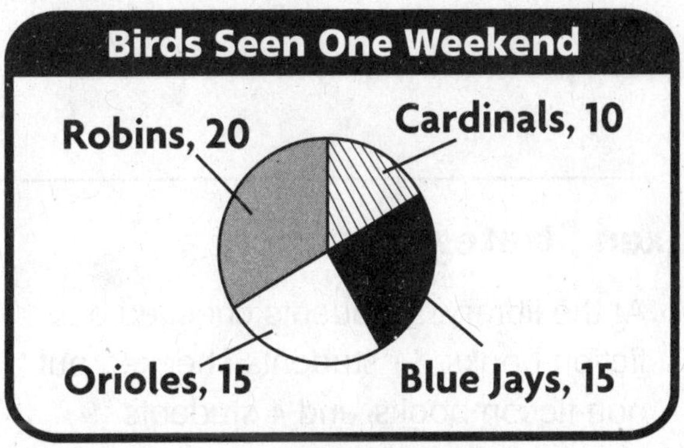

Problem Solving and TAKS Prep

6. **What if** Kendra makes a line graph showing that she exercised 15 minutes on Monday, 22 minutes on Tuesday, 38 minutes on Wednesday, and 35 minutes on Friday. What would the trend of her line graph show?

____

____

7. A circle graph shows that most people prefer walking. It also shows more people prefer biking than swimming. Explain what the circle graph might look like.

____

____

____

Name ____________________

Problem Solving Workshop Strategy: Use Logical Reasoning

Problem Solving Strategy Practice

Draw a Venn diagram to solve.

1. Nine students wrote reports about photosynthesis, 7 students wrote reports about transport tissues in plants, and 3 students wrote about photosynthesis and transport tissues in plants. How many students wrote reports?

____

2. During a free period, 7 students used the computers, 8 students played board games, and 4 students used the computer and played board games. How many students used the computer or played board games during the free period?

____

Mixed Strategy Practice

3. At the library, 5 students checked out fiction books, 11 students checked out non-fiction books, and 4 students checked out fiction and non-fiction books. How many students checked out fiction or non-fiction books from the library?

____

4. During a science experiment, Nora records the following data: day 1 – 14 insects, day 2 – 28 insects; day 3 – 42 insects; day 4 – 56 insects. If the number of insects continues to increase in this way, how many insects will there be on day 8?

____

USE DATA For 5–6, use the table.

5. Hank spent $26.06 on two supplies. Which two supplies did he buy?

____

6. Madison bought the most expensive item. Jerry bought safety goggles and a ruler. How much more did Madison spend than Jerry?

____

Science Supplies Sale

Science Supply	Price
ruler	$2.39
tongs	$11.50
graduated cylinder	$8.71
hand lens	$19.95
safety goggles	$14.56

Name ____________________

Lesson 22.1

Make Pictographs and Bar Graphs

USE DATA For 1–2, use the graphs below.

1. What scale and interval are used in the bar graph?

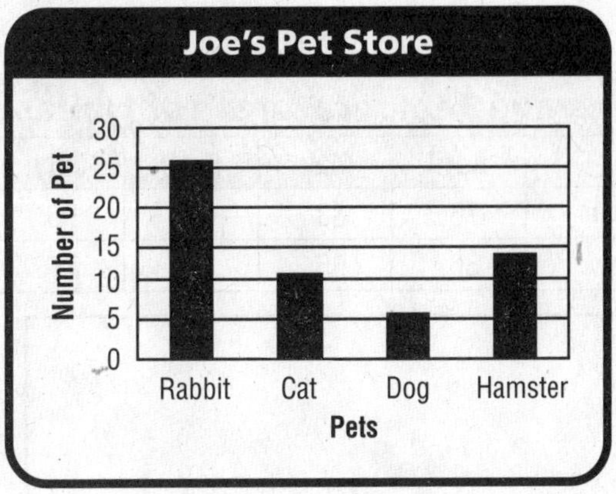

2. Explain how the bars in the graph would change if the interval changed to 10.

3. Make a bar graph for the data.

Favorite Book Types at Woodcreek School	
Book Type	**Boys**
Mystery	35
Fantasy	15
Poetry	10
Sports	40

Problem Solving and TAKS Prep

4. What is a reasonable scale and interval to graph the data in the table below?

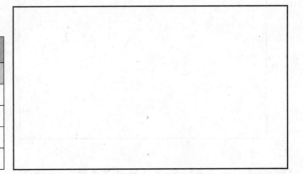

Number of CDs and Movies	
Student	**CDs**
Deb	20
Jose	31
Joanne	10
Mark	32

5. Which interval would you use to make a bar graph of the following data: 60, 55, 40, 35, and 65?

A 2 **C** 10

B 25 **D** 5

6. During soccer sign-up, 9 players joined the first week, 13 joined the second week, and 14 joined the third week. Which type of graph would you use to show the data? Explain why you chose that type of graph.

Name____________________

Lesson 22.2

Make Double-Bar Graphs

Make a double-bar graph for each set of data.

1.

Average Temperatures for Dallas				
	April	May	June	July
High	76	83	92	96
Low	56	65	73	77

2. Students were asked their favorite food at the school cafeteria. Seventy boys and 45 girls chose hamburgers, 52 boys and 93 girls chose pizza, and 61 boys and 49 girls chose chicken tenders.

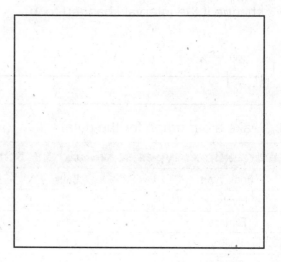

Problem Solving and TAKS Prep

USE DATA For 3–4, use the graph.

3. Jesse collects data to compare the lengths of male and female sharks. Female Tiger sharks are 10 ft long, Great White sharks are 20 ft long, Hammerhead sharks are 12 ft long, and Mako sharks are 9 ft long. Use the data about female sharks to change the bar graph at the right to a double-bar graph.

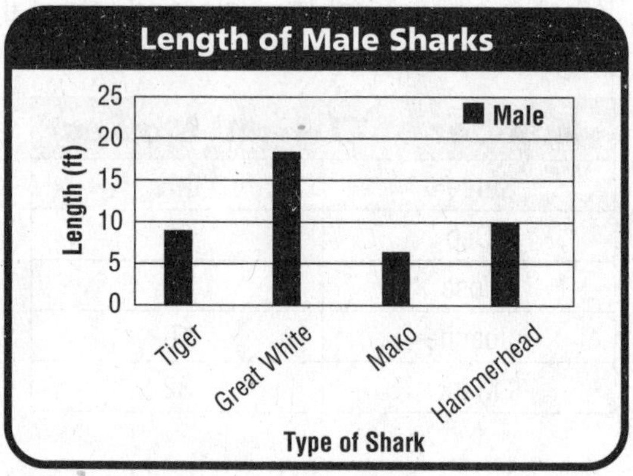

4. Which species of male and female sharks are the most similar in length?

A Tiger **C** Mako
B Great White **D** Hammerhead

5. Which type of graph would best represent the lengths of male and female jellyfish in the Gulf of Mexico?

F Bar Graph **H** Double-Bar Graph
G Line Graph **J** Circle Graph

Name____ **Lesson 22.3**

Make Line Graphs

1. Make a line graph using the information provided.

Total Snowfall on Joey's Birthday					
Time	8 a.m.	11 a.m.	2 p.m.	5 p.m.	8 p.m.
Inches	1	3	5	6	8

2. Celia recorded the weights of her two puppies, Ike and Eli, for 3 months. On the first day, Ike weighed 2 pounds and Eli weighed 2.5 pounds. After one month, Ike weighed 6 pounds and Eli was 5 pounds. After 2 months, Ike was 11 pounds and Eli was 11.5 pounds. After 3 months, Ike weighed 31 pounds and Eli weighed 34 pounds.

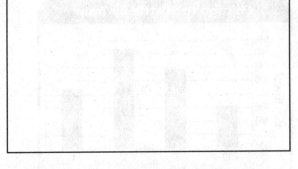

Problem Solving and TAKS Prep

USE DATA For 3–4, use the chart below.

3. Make a line graph for the data in the table below.

Depth of Water in Emma's Pool					
Minute	0	5	10	15	20
Depth (in.)	0	2	6	8	20

4. Between which minutes did the pool gain the most depth?

   A 0–5
   B 5–10
   C 10–15
   D 15–20

5. Make a line graph for the data in the table below.

Tommy's Height				
Age (years)	1	3	5	7
Height (in.)	29	34	37	43

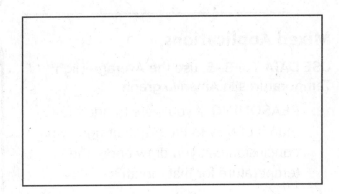

Name____________________ **Lesson 22.4**

Problem Solving Workshop Skill: Draw Conclusions

Problem Solving Skill Practice

Solve problems by using the skill *draw conclusions*.

1. Joseph graphed the number of times the students from Mrs. Nadela's class saw a movie over the summer. Find the range. What conclusion can you draw from the graph?

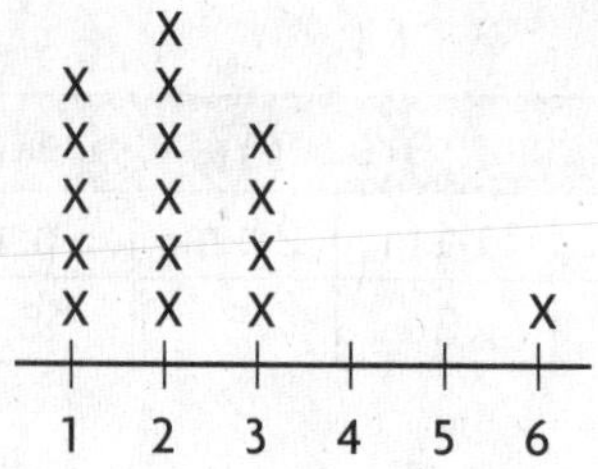

Number of Movies Seen Over the Summer

____

____

2. Sam is traveling from Chicago to Seattle. The graph at the right shows how far the train travels every ten hours. If the trip is 2,206 miles, will Sam arrive in Seattle in 40 hours? Explain why or why not.

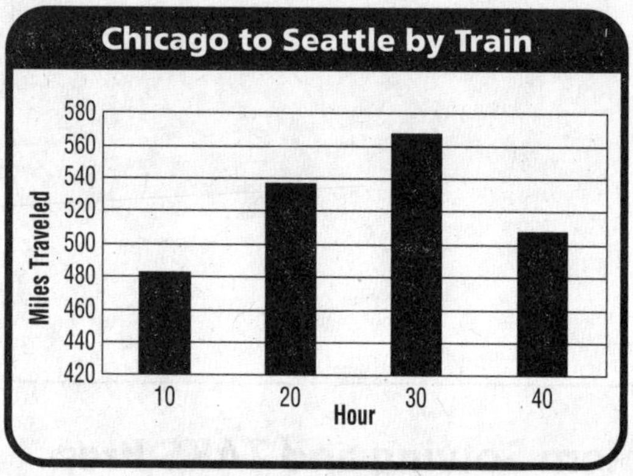

____

____

Mixed Applications

USE DATA For 3–5, use the Average High Temperatures in Amarillo graph.

3. **REASONING** If you were to add the month of May to the graph at right, what conclusion can you draw about the temperature for that month?

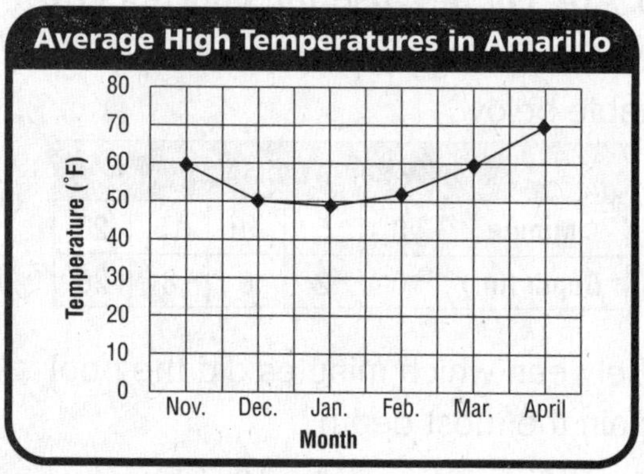

____

____

4. Which months were warmer than 55°F?
   - **A** March and April
   - **B** Nov., March, and April
   - **C** Nov., Feb., March, and April
   - **D** All six months

5. Which statement is false?
   - **F** The average high temperature for Dec. is 50°F.
   - **G** The temperature rose from Jan. to April.
   - **H** Jan. is the coldest month in Amarillo.
   - **J** Nov. and April are the warmest months.

Name____

Choose the Appropriate Graph

Match each graph with the type of data it shows.

A. shows change over time

B. compares data by category

C. uses symbols or pictures

D. compares parts of a group to the whole

1.

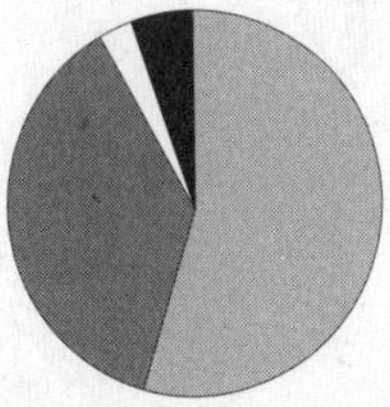

2.

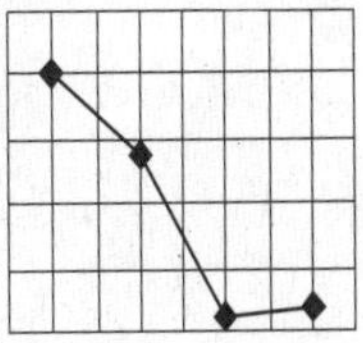

3.

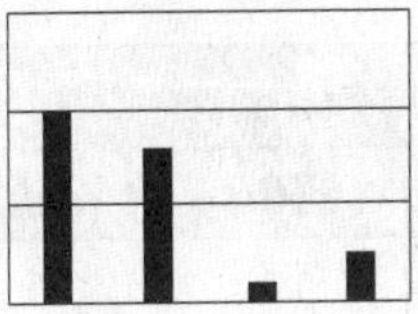

4.

Poodle	• • • •
Black Lab	• • • • • • • •
Beagle	• • •
Boxer	$\frac{1}{2}$

1. ______ 2. ______ 3. ______ 4. ______

Choose the best type of graph or plot for the data. Explain your choice.

5. Types of cars owned by 20 parents

____

____

6. Miles swam by whales per hour

____

____

7. Election results of four candidates

____

____

8. How Victoria spent her lunch hour

____

9. Water evaporated over 10 days

____

10. Hours Raul worked each of the past 6 days

____

Problem Solving and TAKS Prep

11. Draw a graph or plot that best displays the set of data.

Visitors to the Alamo by the Minute

Minute	Visitors
1	14
2	30
3	45
4	65

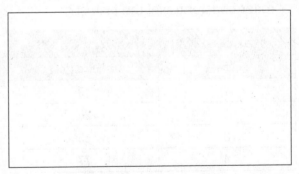

12. You have been given the task of finding the brand of sneakers 15 boys and 15 girls wear. Which graph would best show your results?

A Double-Bar Graph **C** Pictograph

B Line Plot **D** Line Graph

13. Describe a situation in which you would use a circle graph to show data.

____

____

____

Name________________________ Lesson 22.6

Make Graphs

Make the most appropriate graph for each set of data that answers the given question.

1. What fraction represents KMAT listeners?

Radio Station Listeners				
Station	KSOH	KCAH	KTOA	KMAT
Listeners	33	25	11	31

2. How many more students enjoy Galveston than Houston for camping and boating?

Students' Favorite Vacation Spot Activities		
	Galveston	Houston
Sightseeing	26	79
Boating	83	8
Camping	65	40
Shopping	52	59

Problem Solving and TAKS Prep

3. Between which two seconds did the roller coaster travel the fastest?

Distance of Wizard Roller Coaster	
Seconds	Distance (feet)
1	10
2	26
3	65
4	77

____

4. Which type of graph would you make if you want to chart your growth over 12 months?

A Bar Graph
C Pictograph
B Line Graph
D Circle Graph

5. What type of graph is most appropriate if you want to show the number of gold and silver medals won in four different Olympic sports?

____

Name____ Lesson 23.1

Probability Experiments

USE DATA For 1–4, use the table.

1. Rachel pulled a marble from a bag, recorded its color, and put the marble back in the bag. She did this 30 times and recorded her results in the table. Predict how many times out of 80 pulls that Rachel would pull a red marble from the bag.

Marble Experiment				
	Red	Blue	Green	White
Number of pulls	𝍸 I	𝍸 II	𝍸	𝍸 𝍸 II
Total	6	7	5	12

____

2. What is the experimental probability of Rachel pulling

a red marble? _______ a blue marble? _______ a green marble? _______ a white marble? _______

3. **WRITE Math** Based on experimental probabilities, would you predict that Rachel would pull a red or white marble more often if she pulled a marble from the bag 60 more times? Explain.

____

4. Predict the number of times out of 60 pulls that Rachel would pull a red or green marble from the bag.

____

5. How would it compare to your prediction about the number of times out of 60 pulls that Rachel would pull a marble that is not blue or green from the bag?

____

Name________________________________

Describe Outcomes

Use the spinner. Write the probability of each event.
Tell whether the event is certain, likely, unlikely, or impossible.

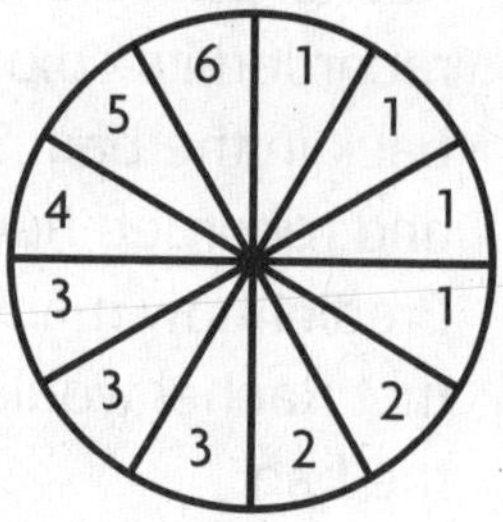

1. spinning 1 ____________

2. spinning 2 ____________

3. spinning 3 ____________

4. spinning 4 ____________

5. spinning 6 ____________

6. spinning 8 ____________

7. spinning 1 or 3 ____________

8. spinning a number less than 7 ____________

9. spinning a number greater than 1 ____________

10. spinning a number other than 10 ____________

11. spinning a number other than 3 ____________

12. spinning a number other than 1, 2, 3, 4, 5, or 6 ____________

Problem Solving and TAKS Prep

USE DATA For 13–16, use the table at the right.

13. What is the sample space for the marbles?

____

Marble Inventory				
Yellow	Red	Blue	Green	White
8	4	2	1	5

14. Reasoning Describe an event whose probability is just as likely as it is unlikely.

____

15. Pull one marble from the bag without looking. Which event is impossible?

A pulling a yellow marble
B pulling a green marble
C pulling a black marble
D pulling a blue marble

16. Pull one marble from the bag without looking. Which event is certain?

F pulling green
G pulling any color but green
H pulling black
J pulling yellow, green red, white, or blue

Name______________________________________ **Lesson 23.3**

Predict Outcomes of Experiments

Use the table to find the experimental probability. Then predict the outcome of future trials.

1. number of wins in 25 more games

Games	
Wins	Losses
卌 卌 ||	卌 |||

____

2. number of yellow in 40 more pulls

Tile Pulls		
Blue	Red	Yellow
卌 卌 ||||	卌 ||	|||

____

3. number of white in 15 more pulls

Marble Pulls			
Red	Black	white	Green
|||	|	卌 ||	卌 卌

____

4. number of losses in 30 more games

Games	
Wins	Losses
卌 卌 卌 |||	||

____

5. number of blue in 48 more pulls

Tile Pulls		
Red	Green	Blue
|||	||	卌 卌 |

____

6. number of red in 40 more pulls

Marble Pulls			
Red	Green	Blue	Yellow
卌 卌	||||	|||	卌 |||

____

Problem Solving and TAKS Prep

7. Without looking Marcus pulls 3 blue marbles, 1 white marble, and 6 red marbles from a bag, and returns them to the bag each time. Predict how many times he will pull a blue or white marble if he pulls a marble from the bag 60 more times.

____

8. Karyn won 5 of her last 6 tennis matches. How many times will she win in 24 more matches?

____

How many times will she win in 42 more matches?

____

9. Samuel drew a blue marble out of a bag in 8 out of 12 pulls. Predict how many times he will draw blue in 30 pulls.

A 16 times
B 18 times
C 20 times
D 26 times

10. Sheeza drew a silver tile out of a bag in 1 out of 4 pulls. Predict how many times she will draw a silver tile in 20 pulls.

F 4 times
G 5 times
H 15 times
J 17 times

Name ______________________________

Tree Diagrams

Use the letter cards or color cards mentioned in the problem.
Draw a tree diagram to find the number of possible outcomes.

A	B	C	D	E	Yellow	Red	Blue	Green

1. draw a vowel from one bag and a consonant from another bag

____

2. draw a color card at random and toss a coin

____

3. draw a letter card at random and toss a coin

____

Problem Solving and TAKS Prep

4. Grace packed a tan, white, and blue shirt to wear with tan, blue, and brown slacks. How many different outfits does he have?

____

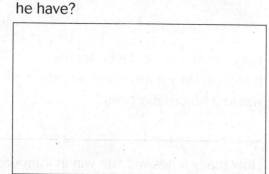

5. Grace packed a blouse, a shirt, and a sweater to wear with jeans, shorts, slacks, and a skirt. How many different outfits does she have?

____

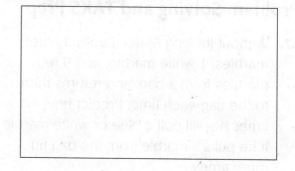

6. Tate has a brown, red, blue, and white shirt. He has brown, blue, and black slacks. How many different ways can he combine shirts and pants?

   A 7
   B 12
   C 14
   D 24

7. Dolly can make a peanut butter, cheese, or vegetable sandwich. She can put it on whole wheat, white, or rye bread. How many sandwiches can she make?

   F 3
   G 6
   H 9
   J 12

Name____ Lesson 23.5

Combinations and Arrangements

Make a list or draw a tree diagram to find the total number of choices.

1. combination, sandwich: cheese, peanut butter, or tuna; whole wheat or rye bread

____

2. combination, pizza: thick or thin crust; cheese, vegetable, or pepperoni topping

____

3. arrangement, ways to place the letters in the word ONE

____

4. arrangement, ways to read a play, a novel, and a short story

____

5. combination, dog food: dry or canned; lamb, beef, or liver

____

6. arrangement, ways that Amy, Bob, and Cleo can wait in a line

____

Problem Solving and TAKS Prep

7. Trent has red, green, and white sweaters. In how many different ways can he wear one sweater on Sunday, Monday, and Tuesday?

____

8. A restaurant offers a choice of rice or potatoes, corn or tomatoes, and chicken or beef with each meal. How many different dinners does it offer?

____

9. How many 3-digit numbers can be made from the digits in the number 251?

   A 3 **C** 5
   B 4 **D** 6

10. How many 4-digit numbers can be made from the digits in the number 4,379?

   F 4 **H** 24
   G 16 **J** 40

Name ____________________

Problem Solving Workshop Strategy: Compare Strategies: Draw a Diagram and Act It Out

Problem Solving Strategy Practice

Draw a diagram or act it out to solve the problem.

1. Danita is playing a word game. She chooses one letter from the word UP and another letter from the word DOWN. How many different combinations of letters could she choose?

____

2. In a game of Toss and Toss, players take turns tossing a coin and tossing a 1 to 6 number cube. The winner is the first to toss every different combination. How many combinations are there?

____

Mixed Applications

3. At the end of a game, Jonathan has 8 more points than Tina. The sum of their scores is 50. What is Tina's score?

____

4. In a game, players toss a coin and spin a spinner with 4 different colors on it. How many different coin and color combinations are there?

____

5. Angelica has a rectangular garden of tulips planted in rows and columns. The tulips are each one foot apart. The garden is 3 feet by 5 feet. How many tulips did she plant?

____

6. Rita, Henry, Celia, and Max are standing in line. Max is between Henry and Rita. Henry is between Max and Celia. Henry ahead of Max. In what order are they standing in line?

____

Name ____________________

Probability Expressed as a Fraction

Use the colors on the cards below to write a fraction for the probability of the event of randomly choosing the card described.

red	blue	violet	red	tan	red	violet	red	blue

1. blue ________
2. red ________
3. tan ________
4. red or violet ________
5. red or blue ________
6. blue or tan ________
7. yellow ________
8. not red ________

Use a number cube labeled 1 to 6 to write a fraction for the probability of the event of tossing each number.

9. 5 ________
10. not 3 ________
11. 2, 3, 4, or 5 ________
12. not 5 or 6 ________

Problem Solving and TAKS Prep

USE DATA For 13–16, use the spinner.

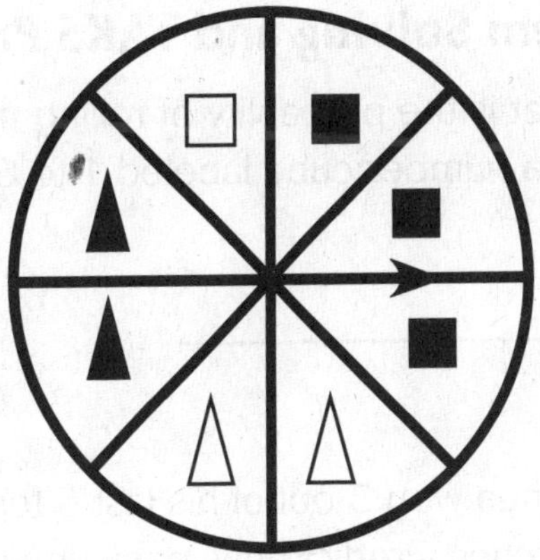

13. What is the probability that the pointer will land on a square?

____

14. What is the probability that the pointer will land on a black square or black triangle?

____

15. What is the probability that the pointer will land on a white shape?

A $\frac{1}{3}$ C $\frac{3}{5}$

B $\frac{2}{3}$ D $\frac{3}{8}$

16. What is the probability that the pointer will land on a black triangle?

F $\frac{1}{2}$ H $\frac{2}{5}$

G $\frac{1}{4}$ J $\frac{1}{8}$

Name ______________________________

Probability and Predictions

Express the experimental probability as a fraction in simplest form. Then predict the outcome of future trials.

1. 5 blue tiles in 15 pulls
   How many blue tiles in the next 9 pulls?
   ____

2. 9 pennies out of 24 coins
   How many pennies in 16 more coins?
   ____

3. 6 yellow marbles in 27 pulls
   How many blue tiles in the next 18 pulls?
   ____

4. 8 heads in 14 coin tosses
   How many heads in 35 more tosses?
   ____

5. 2 losses in 3 games
   How many losses in 9 more games?
   ____

6. 4 wins in 7 games
   How many wins in 21 more games?
   ____

7. 8 blue marbles in 18 pulls
   How many losses in 27 more pulls?
   ____

8. 3 wins in 4 games
   How many wins in 24 more games?
   ____

Problem Solving and TAKS Prep

9. What is the probablity of rolling 4 or a 5 on a number cube labeled 1 to 6?
   ____

10. What is the probablity of rolling an even number on a number cube labeled 1 to 6?
    ____

11. Joshua won 3 out of his last 5 tennis matches. Predict how many times Joshua will win in 10 more matches?

    A 5 times
    B 7 times
    C 4 times
    D 6 times

12. Annie won 15 out of the 20 games she played in a tournament. Predict how many times Annie will win in 12 more games?

    F 9 times
    G 7 times
    H 4 times
    J 24 times

Fairness

For 1–2, perform the experiment by finding the difference rather than the sum or product of the two number cubes. Subtract the smaller number from the larger number. Find the experimental probabilities after 30 tosses of both number cubes. Then, complete the chart of the possible difference outcomes.

1. There are 6 different differences that can be rolled. What are the differences? Write them in the table.

____

2. Perform the experiment. Complete the table.

____

Number Cubes Experiment

Difference	Tally of Times Tossed	Total	Experimental Probability

3. Find the theoretical probability of each difference.

____

____

____

4. What is common to all differences with the greatest theoretical probability?

____

5. How do the theoretical probabilities compare to the experimental probabilities you found for each difference?

____

6. Suppose the experiment is a game with 6 players and each player is assigned a number from 0 to 5. A player scores a point each time his or her difference is tossed.

a. Why is the game unfair?

____

b. How could the scoring be changed to make it fair? Explain.

____

____

Name________________________________

Problem Solving Workshop Skill: Too Much/Too Little Information

Problem Solving Skill Practice

Tell whether each question has too much, too little, or the right amount of information. Answer the question if it can be solved. If you cannot answer the question, describe the information needed to solve it.

1. Latika spent \$6.30 on lunch, \$9.50 on parking, and the rest of the money she spent on carnival rides and snacks. She started with \$50 and has \$8.70 left from the day at the carnival. How much did she spend on rides?

____

2. A spinning cups carnival ride has 6 red, 8 yellow, and 4 blue cups. The remaining cups are green. If $\frac{1}{4}$ of the cups are red and $\frac{1}{2}$ are yellow, how many are green?

____

3. Vinnie collected \$24.60 in ticket sales before 1:00 P.M. Then 22 people purchased rollercoaster tickets at \$1.20 each and 12 people purchased bumper car tickets at \$1.50 each. How many bumper car tickets did Vinnie sell before 1:00 P.M.?

____

4. Vinnie sells tickets for the rollercoaster and bumper car rides. He collected \$24.60 in ticket sales before 1:00 P.M. Then he sold 22 rollercoaster tickets at \$1.20 each and 12 bumper car tickets at \$1.50 each. How much did Vinnie have in sales then?

____

Mixed Applications

5. **Pose a Problem** Change at least one number in Problem 2, above. Describe the change. Then solve the new problem.

____

____

6. Jessie wants to save 1¢ the first day, 2¢ the second day, 3¢ the third day, 4¢ the fourth day, and so on, adding 1¢ to the amount she saves each day. How much will she have saved in one year of 365 days?

____

SPIRAL REVIEW

Name________________________________ **Week 1**

Spiral Review

For 1–4, write the value of the underlined digit.

1. 2.<u>6</u>5 ____________

2. 12.8<u>1</u> ____________

3. 5.<u>9</u>7 ____________

4. 3.4<u>9</u> ____________

For 5–6, write each number in two other forms.

5. 6.35

____

____

6. two and fourteen hundredths

____

____

For 7–9, find the elapsed time.

7. start: 11:15 A.M.

end: 2:00 P.M. ____________

8. start: 3:30 P.M.

end: 6:45 P.M. ____________

9. start: 9:30 P.M.

end: 4:15 A.M. ____________

For 10–11, find the ending time.

10. start: 4:00 P.M.

elapsed time: 5 hr 15 min ____________

11. start: 10:30 P.M.

elapsed time: 2 hr 20 min ____________

For 12–13, place the numbers where they belong in the Venn diagram.

12.

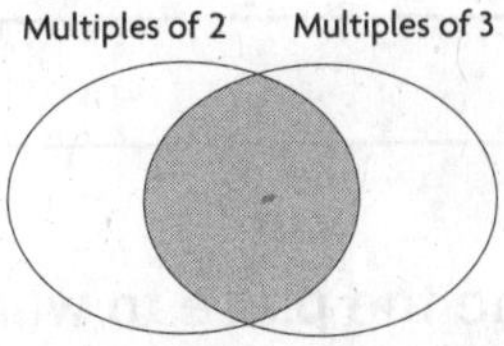

2, 6, 3, 9, 12, 4, 15, 18, 21

13.

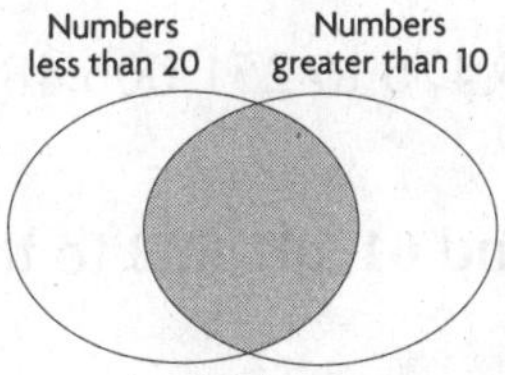

23, 18, 6, 25, 8, 16, 37, 9, 11

For 14–15, tell whether the two figures are *congruent* or *not congruent*.

14.

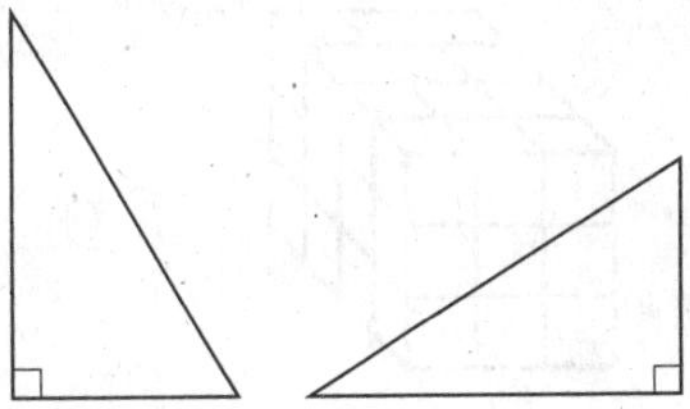

____

15.

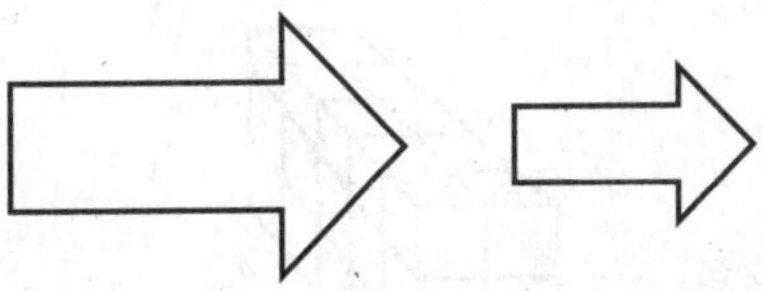

____

Spiral Review

For 1–3, round each number to the place of the underlined digit.

1. 124,516,228 ______
2. 6,732 ______
3. 25,019 ______

For 4-5, name the place to which each number was rounded.

4. 76,812 to 80,000 ______
5. 251,006,475 to 251,006,480 ______

For 6–7, round 61,201,983 to the place named.

6. hundred thousands ______
7. tens ______

Make an organized list to solve.

10. Ken is making tickets for the fair. Each type of ticket will be a different color. There will be adult and child tickets. There will be 1-day, 2-day, and weekly tickets. How many different ticket colors will there be?

______

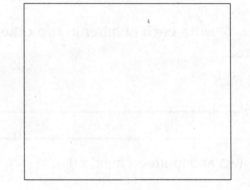

For 8–9, count or multiply to find the volume.

8.

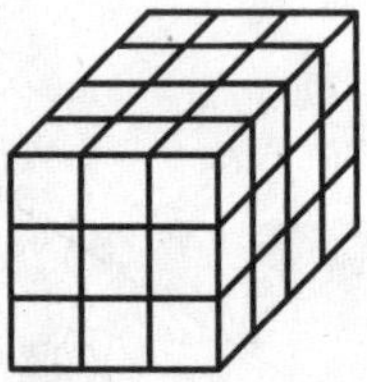

______

9.

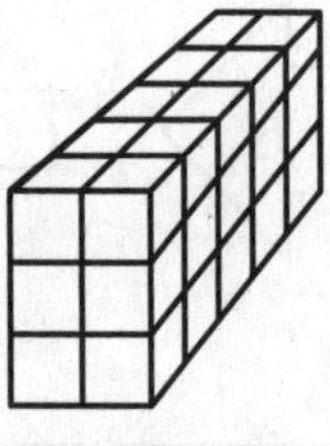

______

For 11–14, find the product.

11. 8×3 ______
12. 4×5 ______
13. 2×7 ______
14. 9×6 ______

For 15–18, find the value of $p \times 6$ for each value of p.

15. $p = 3$ ______
16. $p = 5$ ______
17. $p = 4$ ______
18. $p = 7$ ______

Name ____________________ **Week 3**

Spiral Review

Find the sum or difference.

1. $\begin{array}{r} 91.47 \\ +\ 23.76 \\ \hline \end{array}$

2. $\begin{array}{r} 105.308 \\ -\ 61.487 \\ \hline \end{array}$

3. $\begin{array}{r} 8.759 \\ +\ 5.413 \\ \hline \end{array}$

4. $\begin{array}{r} 2.704 \\ -\ 0.285 \\ \hline \end{array}$

5. $\begin{array}{r} 0.42 \\ 0.309 \\ +\ 2.695 \\ \hline \end{array}$

6. $\begin{array}{r} 18.751 \\ 6.049 \\ +\ 12.201 \\ \hline \end{array}$

For 7–10, find the perimeter of each figure.

7. 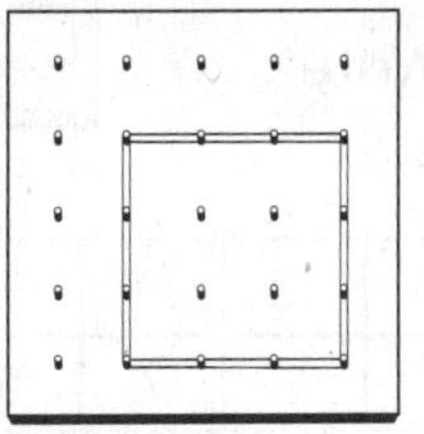
______

8. 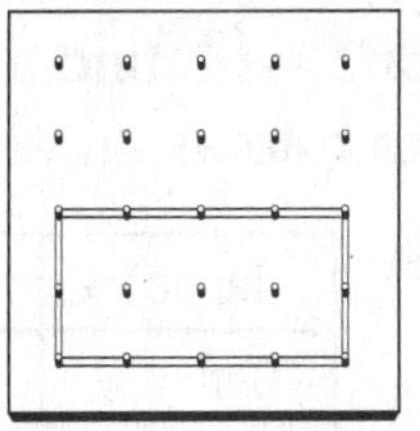
______

9. 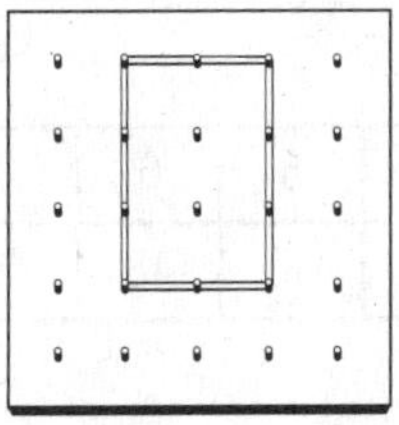
______

10. 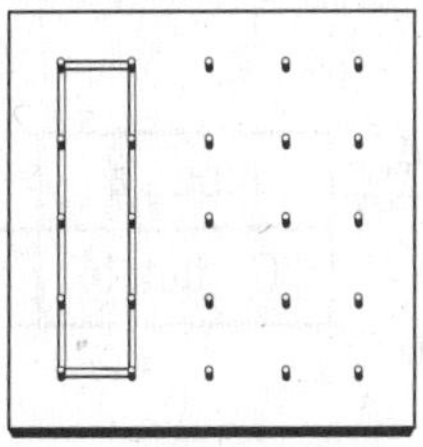
______

For 11–13, use the double-bar graph.

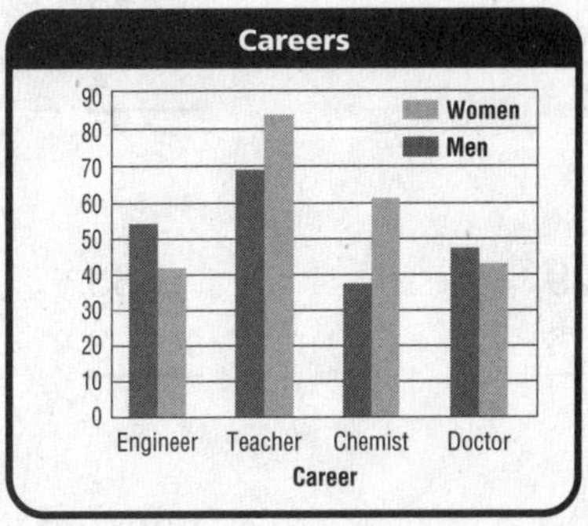

11. What two sets of data are compared in the graph? ______

12. Which careers have more men than women? ______

13. Which career is least popular for women? ______

For 14–17, name the any relationships you see in each figure. Write *intersecting, parallel*, or *perpendicular*. Tell if the figure has *acute, right, obtuse*, or *straight* angles.

14. 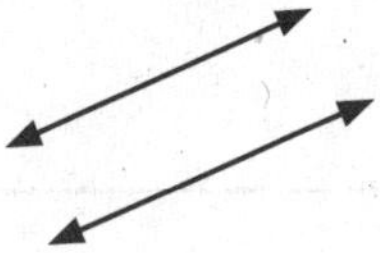
______
______

15. 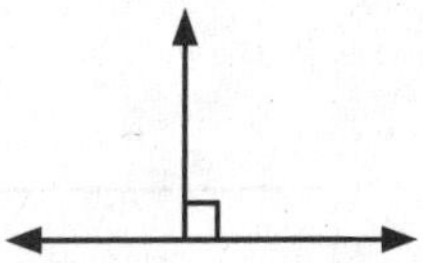
______
______

16. 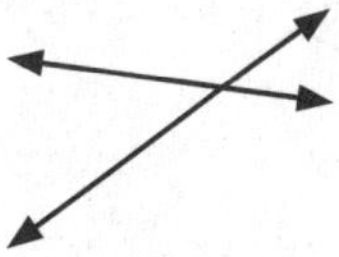
______
______

17. 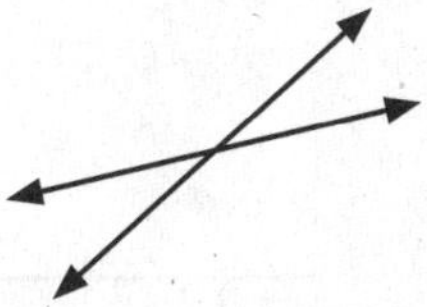
______
______

Spiral Review

For 1–8, estimate. Then find the product.

1. $\begin{array}{r} 26 \\ \times\ 7 \\ \hline \end{array}$

2. $\begin{array}{r} 672 \\ \times\ \ 4 \\ \hline \end{array}$

3. $\begin{array}{r} 429 \\ \times\ \ 6 \\ \hline \end{array}$

4. $\begin{array}{r} 783 \\ \times\ \ 3 \\ \hline \end{array}$

5. 542×5 ______

6. 239×7 ______

7. 3×462 ______

8. 289×6 ______

For 9–10, use the thermometer to find the temperature in °F.

9.

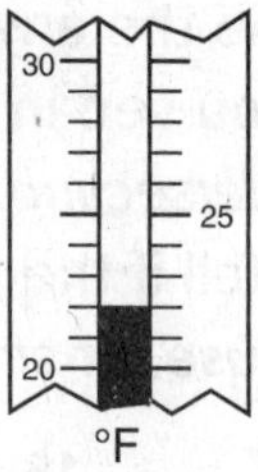

____

10.

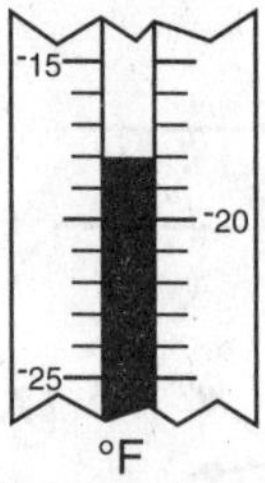

____

For 11–12, use the Favorite Type of Music frequency table. Tell whether each statement is true or false. Explain.

Favorite Type of Music

Type of Music	Votes
Country	43
Rock	37
Rap	34

11. More people chose rap than rock as their favorite.

____

____

12. Country music was chosen by more people than rock.

____

____

For 13–14, find a rule. Write the rule as an equation.

13.

Input, x	9	15	18	21	27
Output, y	3	5	6	7	9

____

14.

Input, a	2	3	5	6	8
Output, b	16	24	40	48	64

____

Spiral Review

For 1–6, divide.

1. $8\overline{)512}$
2. $4\overline{)385}$
3. $5\overline{)247}$
4. $3\overline{)844}$
5. $821 \div 6 =$ ______
6. $198 \div 2 =$ ______

Make a table to solve.

9. How many combinations of two names can be made from the names Sarah, Mark, and Ryan?

____

For 7–8, find the perimeter.

7.

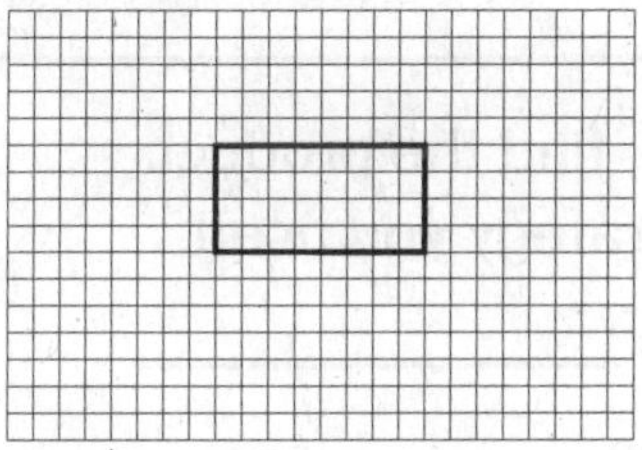

____

8.

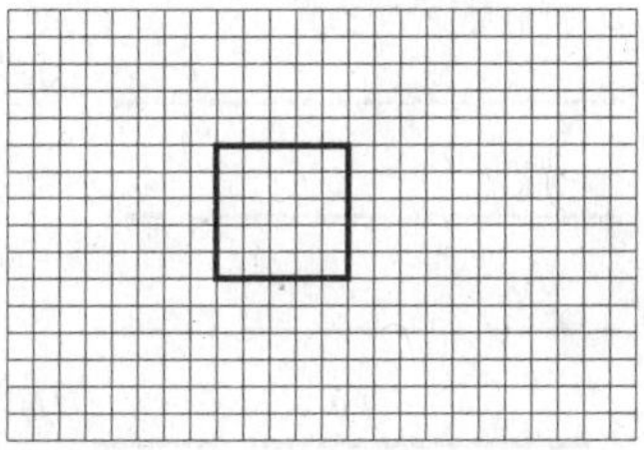

____

For 10–13, classify each figure in as many of the following ways as possible. Write *quadrilateral*, *parallelogram*, *rhombus*, *rectangle*, *square*, or *trapezoid*.

10.

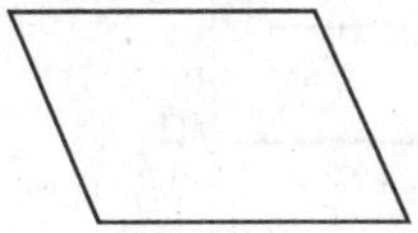

____

____

11.

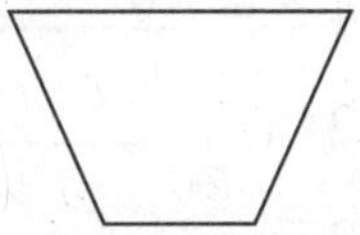

____

____

12.

____

____

13.

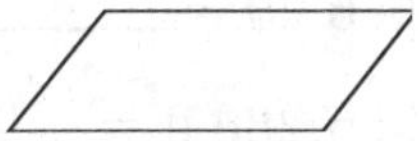

____

____

Spiral Review

For 1–4, use basic facts and patterns to find the missing quotient.

1. 30 ÷ 10 ________
2. 540 ÷ 90 ________
3. 420 ÷ 6 ________
4. $150 ÷ 30 ________

For 5–8, divide. Check your answer.

5. 32)426
6. 47)529
7. 21)741
8. 74)647

For 9–20, change each unit.

9. 24 in. = ________ ft
10. 4 c = ________ pt
11. 24 ft = ________ yd
12. 2 T = ________ lb
13. 2 c = ________ fl oz
14. 2 gal = ________ qt
15. 6 yd = ________ ft
16. 5,280 ft = ________ mi
17. 4 ft = ________ in.
18. 3 lb = ________ oz
19. 15 ft = ________ yd
20. 2,000 lb = ________ T

For 21–24, use the graph below.

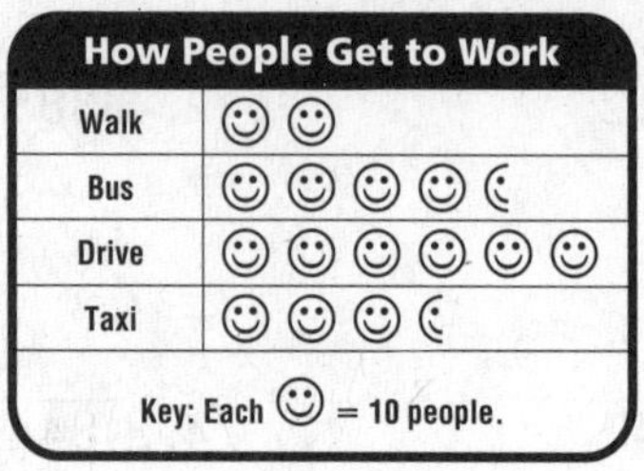

21. How many people take the bus to work? ________
22. How would you change the graph if 25 people walk to work?

____

____

23. How many people were surveyed in all? ________
24. How many more people drive than take a taxi to work? ________

For 25–36, find the product. Tell the strategy you used.

25. 2 × 3 ________
26. 5 × 10 ________
27. 6 × 2 ________
28. 4 × 9 ________
29. 8 × 7 ________
30. 9 × 5 ________
31. 10 × 3 ________
32. 8 × 6 ________
33. 10 × 9 ________
34. 7 × 5 ________
35. 5 × 8 ________
36. 10 × 6 ________

Spiral Review

For 1–4, complete to find the sum or difference.

1.
```
   54,639
 − 37,840
 1__,7__9
```

2.
```
   738,521
 + 601,994
 1,34__,__1__
```

3.
```
   4,193
 + 5,570
 __,7__3
```

4.
```
   65,574
 −  7,321
 5__,2__3
```

For 5–6, estimate. Then find the sum or difference.

5.
```
   84,679
 − 39,213
```

6.
```
   5,807,436
 + 2,789,015
```

Make a tree diagram to find the number of possible combinations.

8. Shirt choices
color: blue, green, white
style: long-sleeve, short-sleeve

Find the perimeter and area of the figure. Then draw another figure that has the same area, but a different perimeter.

7.

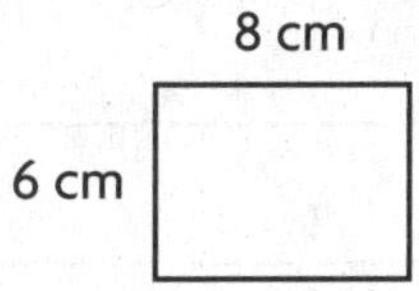

For 9–10, tell whether the figure appears to have *line symmetry*, *rotational symmetry*, *both*, or *neither*.

9. 

10. 

For 11–12, draw the line or lines of symmetry.

11. 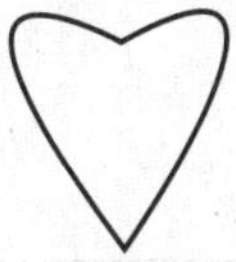

12. 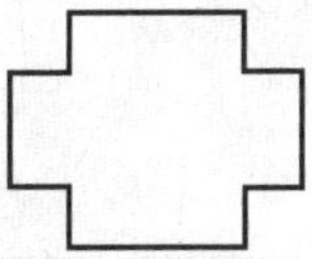

Spiral Review

For 1–11, write the common factors for each pair of numbers.

1. 24, 40 ____________
2. 36, 60 ____________
3. 9, 27 ____________
4. 15, 30 ____________
5. 18, 42 ____________
6. 8, 40 ____________
7. 14, 56 ____________
8. 21, 42 ____________
9. 6, 15 ____________
10. 12, 30 ____________
11. 10, 45 ____________

For 14–18, choose 5, 10, or 100 as the most reasonable interval for each set of data. Explain your choice.

14. 90, 350, 260, 185, 415 ____________
15. 7, 23, 25, 18, 11 ____________
16. 52, 76, 24, 54, 61 ____________
17. 218, 371, 882, 119, 505 ____________
18. 10, 20, 14, 7, 17 ____________

For 12–13, count or multiply to find the volume.

12.

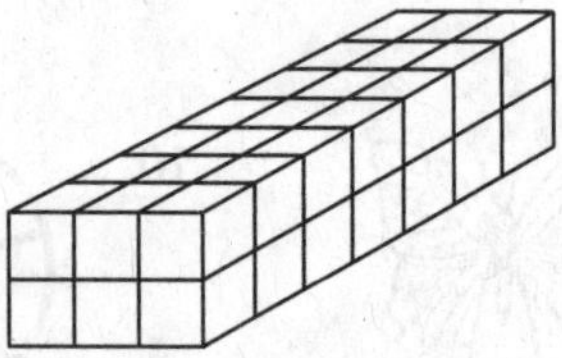

____

13.

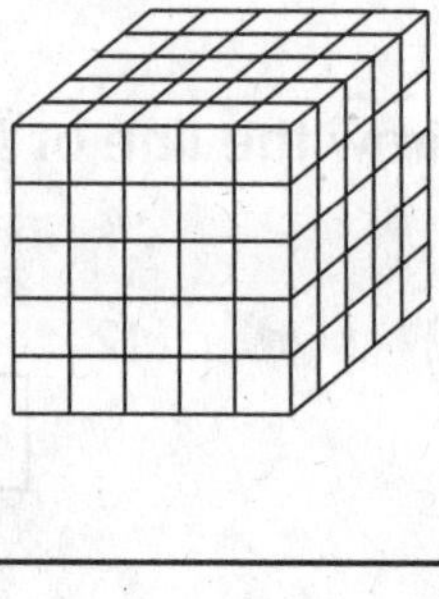

____

For 19–24, use counters to show all arrays for each number. Write prime or composite.

19. 35 ____________
20. 9 ____________
21. 29 ____________
22. 101 ____________
23. 75 ____________
24. 55 ____________

Spiral Review

For 1–6, compare.

Write <, >, or = for each ◯.

1. $\frac{1}{3} \bigcirc \frac{1}{2}$
2. $\frac{5}{7} \bigcirc \frac{3}{5}$
3. $4\frac{3}{7} \bigcirc 4\frac{2}{5}$
4. $3\frac{1}{3} \bigcirc 3\frac{4}{12}$
5. $2\frac{7}{12} \bigcirc 2\frac{5}{8}$
6. $2\frac{2}{3} \bigcirc 1\frac{8}{15}$

For 7–8, write in order from least to greatest.

7. $\frac{1}{3}, \frac{5}{6}, \frac{1}{6}$ ____________

8. $2\frac{5}{6}, 3\frac{2}{3}, 2\frac{4}{9}$ ____________

For 9–10, find the perimeter and area of each figure.

9.

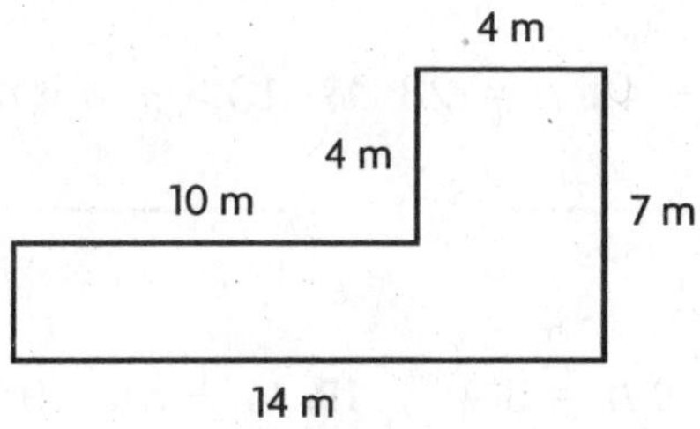

____

10.

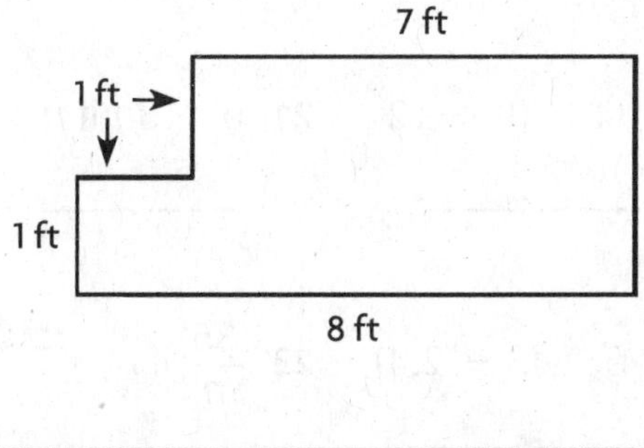

____

Make a model to find all possible combinations.

11. Dance Class choices
time: morning, afternoon
day: Tuesday, Wednesday, Thursday
teacher: Rosa, Mark

____

For 12–14, draw circle *A* with a 4-centimeter radius. Label each of the following.

12. radius: $\overline{BA}$

13. chord: $\overline{CD}$

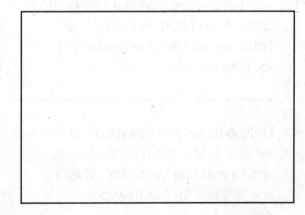

14. diameter: $\overline{FG}$

Name ____________________ **Week 10**

Spiral Review

For 1–10, find the sum. Write it in simplest form.

1. $\frac{1}{3} + \frac{1}{3} =$ ____

2. $\frac{1}{6} + \frac{3}{6} =$ ____

3. $\frac{4}{9} + \frac{1}{9} =$ ____

4. $\frac{3}{7} + \frac{1}{7} =$ ____

5. $\frac{5}{12} + \frac{3}{12} =$ ____

6. $\frac{3}{4} + \frac{1}{4} =$ ____

7. $\frac{3}{10} + \frac{6}{10} =$ ____

8. $\frac{5}{9} + \frac{2}{9} =$ ____

9. $\frac{1}{6} + \frac{2}{6} =$ ____

10. $\frac{4}{5} + \frac{1}{5} =$ ____

For 14–15, make a table to solve.

14. How many combinations of two sports can be made with the sports soccer, baseball, and lacrosse?

____

15. How many combinations of 3 letters can be made from the word RED?

____

For 11–13, use a calendar to solve.

11. The zoo will be offering discount tickets from January 3 to January 29. How many days will tickets be discounted?

____

12. The pet store is having a sale on dog and cat food from February 1 to February 16. How many days will the food be on sale?

____

13. Delia paid for her newspaper delivery on July 1. She last paid for it three weeks and four days ago. When did she last pay for her newspaper delivery?

____

For 16–23, evaluate each expression.

16. $n \div 4$ if $n = 28$ ____

17. $13.8 - n$ if $n = 7$ ____

18. $6n$ if $n = 3$ ____

19. $(4 + n) - 9$ if $n = 6$ ____

20. $(4 \times 9) \div 12$ ____

21. $6 + 3n$ if $n = 3$ ____

22. $5 \times (3 - 2.4)$ ____

23. $\frac{35}{n}$ if $n = 7$ ____

Spiral Review

For 1–8, estimate. Then find the sum or difference. Write it in simplest form.

1. $\begin{array}{r} 6\frac{7}{10} \\ -\ 1\frac{6}{10} \\ \hline \end{array}$

2. $\begin{array}{r} 8\frac{3}{12} \\ -\ 3\frac{1}{12} \\ \hline \end{array}$

3. $5\frac{1}{3} + 7\frac{2}{3} =$ ____

4. $9\frac{5}{9} - 2\frac{2}{9} =$ ____

5. $4\frac{1}{8} + 3\frac{5}{8} =$ ____

6. $3\frac{1}{4} + 6\frac{2}{4} =$ ____

7. $8\frac{3}{8} + 4\frac{1}{8} =$ ____

8. $8\frac{7}{8} - 8\frac{3}{8} =$ ____

Make a tree diagram to find the number of possible combinations.

14. Outfit choices
pants: blue, tan, jeans
shirt: sweater, button-down, T-shirt

____

For 9–11, find the elapsed time.

9. start: 10:45 A.M.
end: 1:00 P.M. ____________

10. start: 4:30 P.M.
end: 7:15 P.M. ____________

11. start: 8:30 P.M.
end: 10:45 P.M. ____________

For 12–13, find the ending time.

12. start: 3:00 P.M.
elapsed time: 4 hr 20 min ____________

13. start: 10:30 A.M.
elapsed time: 3 hr 45 min ____________

For 15–19, name a solid figure that is described.

15. one circular face ____________

16. all triangular faces ____________

17. 5 faces ____________

18. 12 edges ____________

19. 4 vertices ____________

For 20–21, tell whether the net would make a cube. Write *yes* or *no*.

20.

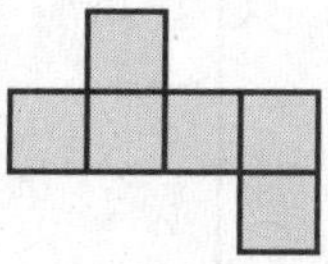

____

21.

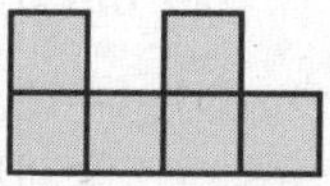

____

Name________________________ **Week 12**

Spiral Review

For 1–6, write each fraction as a decimal.

1. $\frac{3}{5}$ ________
2. $\frac{5}{25}$ ________
3. $\frac{4}{10}$ ________
4. $\frac{37}{100}$ ________
5. $\frac{28}{50}$ ________
6. $\frac{2}{100}$ ________

For 7–12, write each decimal as a fraction in simplest form.

7. 0.6 ____
8. 0.45 ____
9. 0.26 ____
10. 0.52 ____
11. 0.34 ____
12. 0.35 ____

For 15–19, choose 5, 10, or 100 as the most reasonable interval for each set of data. Explain your choice.

15. 35, 21, 47, 72, 23

____

____

16. 4, 18, 16, 7, 13

____

____

17. 88, 275, 119, 458, 605

____

____

18. 12, 15, 21, 17, 24

____

____

19. 19, 33, 62, 47, 34

____

____

For 13–14, find the area.

13.

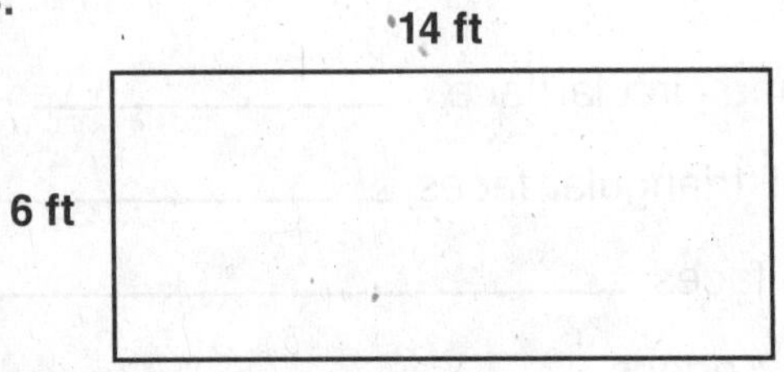

____

14.

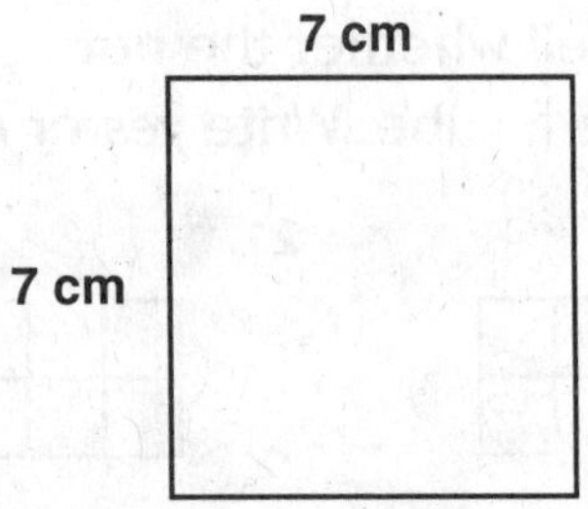

____

For 20–22, write an equation. Tell what the variable represents.

20. Brad has 28 oranges. He gives some away. He now has 11 oranges. How many oranges does Brad give away?

____

____

21. Hank has some stamps. He uses 14 of them. He has 17 stamps left. How many stamps did Hank originally have?

____

____

22. Gina divides some crackers among her 4 friends. She gives each friend 6 crackers. How many crackers did Gina have?

____

____

Spiral Review

For 1–4, write an equivalent fraction.

1. $\frac{4}{6}$ ______ 2. $\frac{2}{5}$ ______

3. $\frac{8}{12}$ ______ 4. $\frac{4}{16}$ ______

For 5–8, tell which fraction is not equivalent to the others.

5. $\frac{6}{9}, \frac{3}{4}, \frac{8}{12}$ ______

6. $\frac{2}{6}, \frac{3}{9}, \frac{4}{10}$ ______

7. $\frac{1}{2}, \frac{3}{7}, \frac{4}{8}$ ______

8. $\frac{5}{15}, \frac{6}{10}, \frac{3}{5}$ ______

For 11–13, use the tally table.

Family Vacation Survey		
Days	Families	Frequency
5	///	
10	卌 ///	
15	卌 卌 //	
20	卌 ////	

11. Complete the frequency column in the table.

12. How many family vacations last 10 days? ______

13. How many family vacation days have the greatest frequency? ______

For 9–10, count or multiply to find the volume.

9.

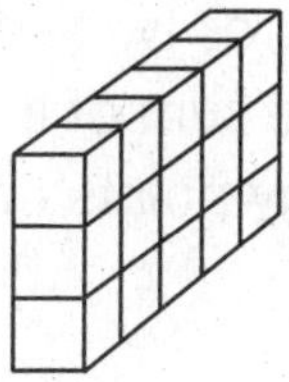

____

10.

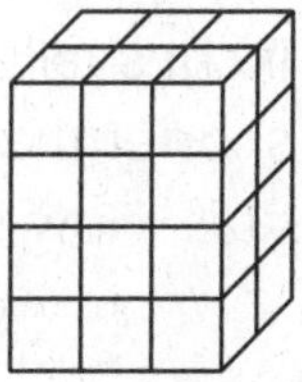

____

For 14–17, write *parallel*, *intersecting*, or *perpendicular* for each.

14.

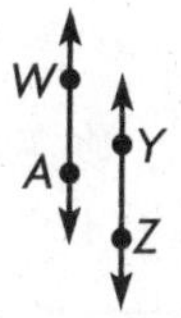

______

15.

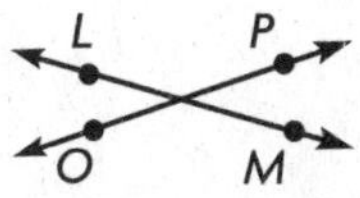

______

16.

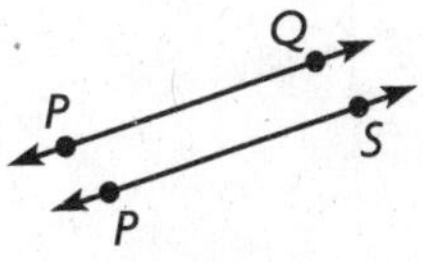

______

17.

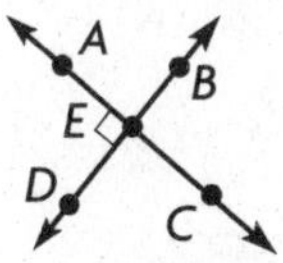

______

Name____ **Week 14**

Spiral Review

For 1–2, order from greatest to least.

1. 23,412; 23,572; 23,012

____

2. 697,581; 699,815; 69,705

____

For 3–5, compare. Write <, >, or = for each.

3. 67,809 ◯ 67,890

4. 19,987 ◯ 19,897

5. 6,521,303 ◯ 6,521,330

For 6–7, order from least to greatest.

6. 2,321,503; 23,205; 231,305

____

7. 64,879; 64,867; 64,789

____

For 8–9, use the thermometer to find the temperature in °C.

8.

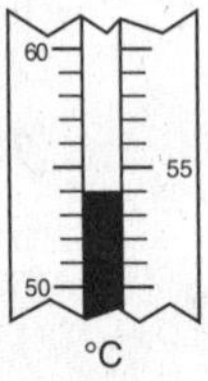

____

9.

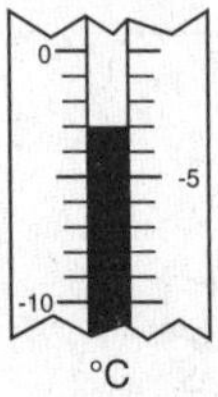

____

For 10–11, make a table to solve.

10. How many combinations of two letters can be made from the letters in the word CUTE? ______________

11. How many combinations of three animals can be made from lion, bear, goose, fox, and wolf? ______________

For 12–13, draw a diagram to solve.

12. Steve scores twice as many points as Brian. The sum of their points is 48. How many points does Brian score?

____

13. Jess is 6 inches taller than Joe. Joe is 4 inches shorter than Karen. Jess is 63 inches tall. How tall is Karen?

____

Spiral Review

For 1–4, write each mixed number as a fraction.

1. $1\frac{1}{3}$ ______ 2. $3\frac{1}{2}$ ______

3. $4\frac{2}{5}$ ______ 4. $2\frac{4}{7}$ ______

For 5–8, write each fraction as a mixed number.

5. $\frac{8}{3}$ ______ 6. $\frac{13}{10}$ ______

7. $\frac{17}{4}$ ______ 8. $\frac{12}{5}$ ______

For 10–12, use the bar graph.

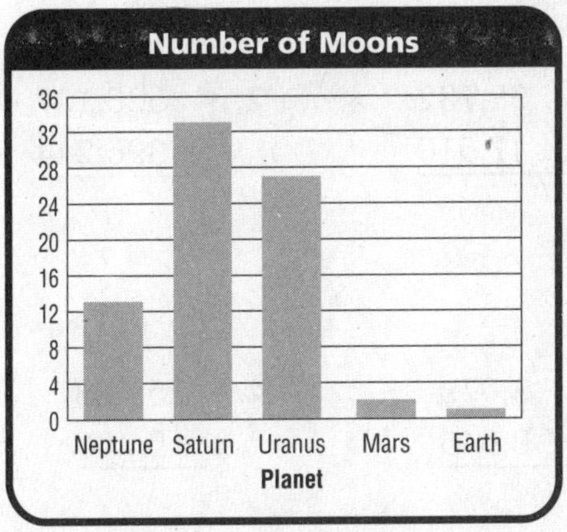

10. Which planet has the greatest number of moons? ______

11. Which planet has 1 more moon than Earth? ______

12. How many moons does Neptune have? ______

Find the perimeter and area of the figure. Then draw another figure that has the same perimeter but a different area.

9.

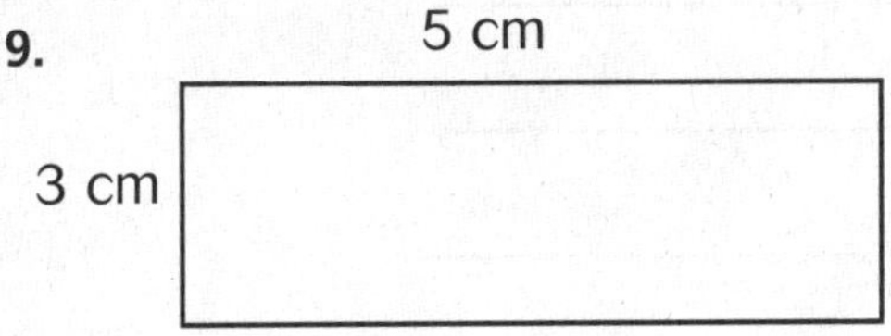

____

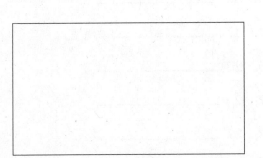

For 13–18, graph and label the following points on the coordinate grid.

13. *A* (4,3) 14. *B* (2,5) 15. *C* (0,7)

16. *D* (3,4) 17. *E* (6,4) 18. *F* (5, 1)

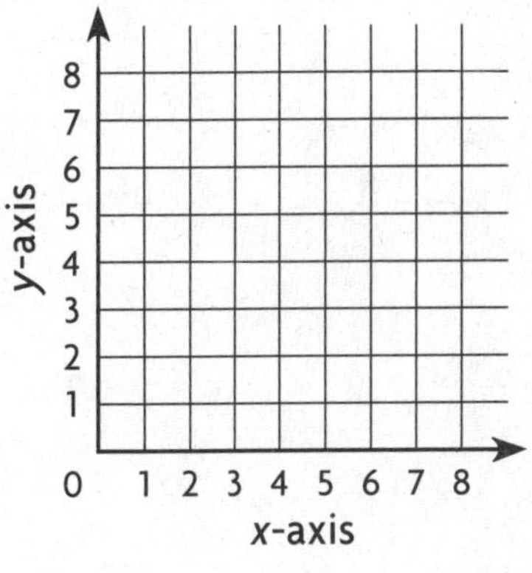

Name________________________________ **Week 16**

Spiral Review

For 1-4, estimate. Then find the sum or difference.

1.
```
  68,732
– 47,510
```

2.
```
  528,137
+ 396,244
```

3.
```
  3,079
+ 8,645
```

4.
```
  25,683
–  6,409
```

For 5-6, complete to find the sum or difference.

5.
```
  61,870
– 35,149
 _6,_2_
```

6.
```
  8,643,572
+ 7,580,421
1_,_2_,_93
```

For 7–16, change the unit.

7. 36 in. = ___ ft

8. 28 qt = ___ gal

9. 5 lb = ___ oz

10. 24 ft = ___ yd

11. 4 pt = ___ fl oz

12. 3 T = ___ lb

13. 3 mi = ___ ft

14. 36 qt = ___ gal

15. 48 c = ___ qt

16. 2.5 T = _____ lb

For 17–19, use the graph below.

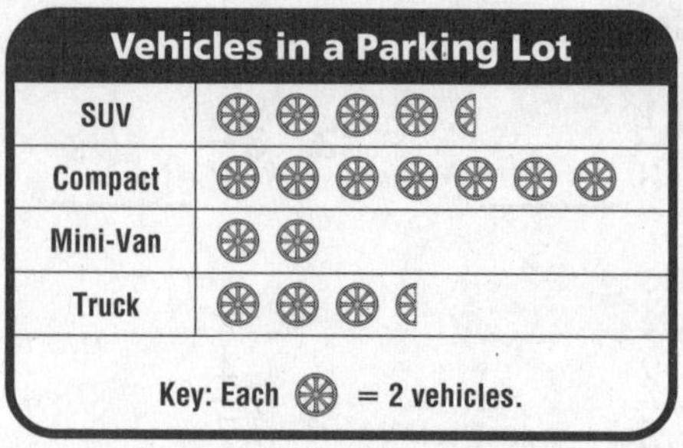

17. How would you change the graph if 15 compact cars were parked in the parking lot?

____

____

18. How many more SUVs than mini-vans are parked in the parking lot?

____

19. How many vehicles are in the parking lot?

____

For 20-29, tell whether adding 2 to each prime number will result in another prime number.

20. 3 ____________

21. 7 ____________

22. 11 ____________

23. 17 ____________

24. 53 ____________

25. 47 ____________

26. 23 ____________

27. 29 ____________

28. 37 ____________

29. 61 ____________

Name____________________ **Week 17**

Spiral Review

For 1–6, compare.
Write <, >, or = for each $\bigcirc$.

1. $\frac{5}{7} \bigcirc \frac{2}{3}$
2. $\frac{4}{5} \bigcirc \frac{6}{7}$
3. $3\frac{1}{5} \bigcirc 3\frac{1}{3}$
4. $1\frac{4}{6} \bigcirc 1\frac{2}{3}$
5. $3\frac{3}{4} \bigcirc 3\frac{7}{12}$
6. $2\frac{1}{2} \bigcirc 2\frac{5}{6}$

For 7–8, write in order from least to greatest.

7. $\frac{5}{6}, \frac{1}{12}, \frac{2}{5}$

____

8. $3\frac{3}{4}, 3\frac{5}{9}, 3\frac{1}{3}$

____

For 9–19, change the unit.

9. 5,000 m = ___ km
10. 8 kL = ___ L
11. 16 m = ___ cm
12. 1,000 mL = ___ metric cups
13. 36 cm = ___ mm
14. 8 metric cups = ___ L
15. 200 cm = ___ m
16. 6,000 L = ___ kL
17. 71 km = ___ m
18. 85 L = ___ mL
19. 620 mm = ___ cm

Make a model to find all possible combinations.

20. Sandwich choices
meat: ham, turkey, roast beef
cheese: American, cheddar
bread: wheat, white

____
____
____
____
____
____
____
____
____
____
____
____

For 21–24, classify each solid figure. Write *prism, pyramid, cylinder, cone,* or *sphere*.

21.
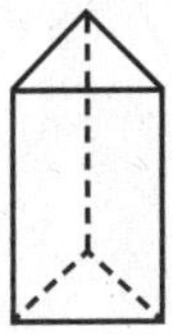

____

22.
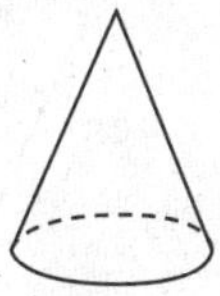

____

23.
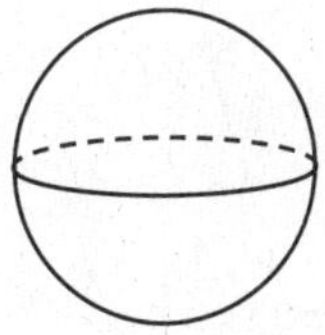

____

24.
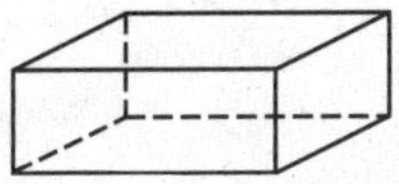

____

Name________________________________ **Week 18**

Spiral Review

For 1–5, write an equivalent fraction.

1. $\frac{1}{2}$ ______ 2. $\frac{3}{9}$ ______

3. $\frac{4}{10}$ ______ 4. $\frac{3}{15}$ ______ 5. $\frac{12}{18}$ ______

For 6–9, tell which fraction is *not* equivalent to the others.

6. $\frac{2}{5}, \frac{4}{10}, \frac{3}{8}$ ______ 7. $\frac{5}{12}, \frac{4}{8}, \frac{3}{6}$ ______

8. $\frac{1}{3}, \frac{5}{9}, \frac{2}{6}$ ______ 9. $\frac{6}{8}, \frac{4}{6}, \frac{9}{12}$ ______

Make a tree diagram to find the number of possible combinations.

12. Activity choices
activity: zoo, park, museum
time: morning, afternoon, evening

____

For 10–11, find the perimeter of each polygon.

10.

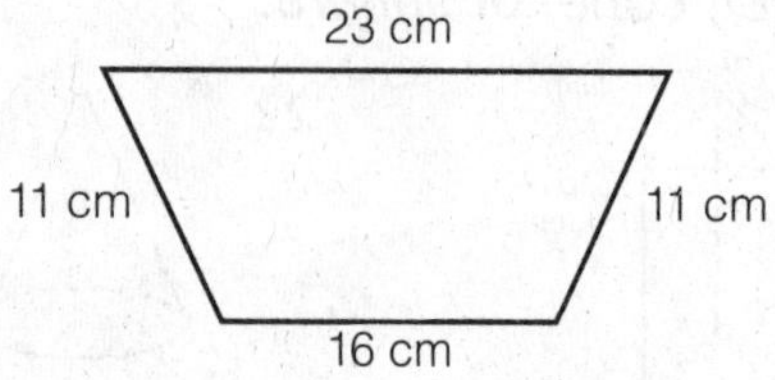

____

11.

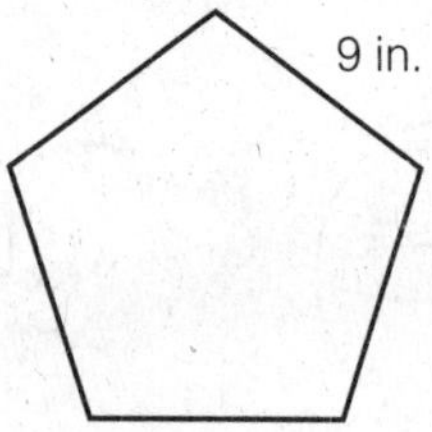

____

For 13–15, find a rule to Write the rule as an equation.

13.

input, x	24	18	12	6
output, y	8	6	4	2

____

14.

input, x	2	5	6	8
output, y	4	10	12	16

____

15.

input, x	20	16	14	12
output, y	11	9	8	7

____

Spiral Review

Find the product. Choose mental math, paper and pencil, or a calculator.

1. 308×52

2. 649×37

3. 582×41

4. 825×24

5. 264×83 ________

6. 68×719 ________

For 9–10, place the numbers where they belong in the Venn diagram.

9.

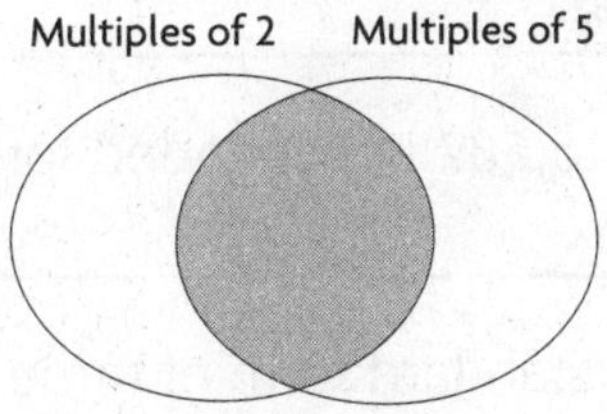

10, 4, 20, 15, 18, 25, 40, 14

10.

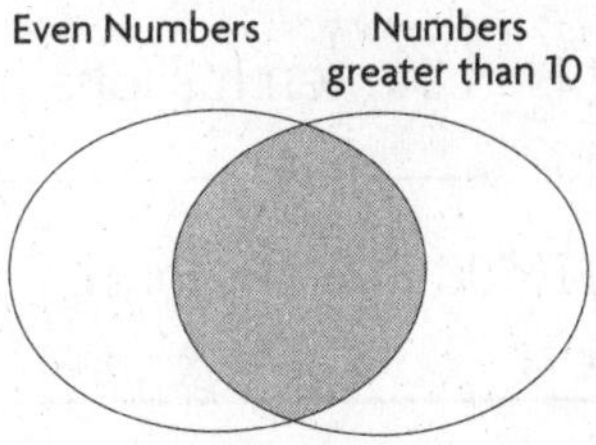

8, 17, 12, 2, 13, 21, 14, 5

For 7–8, find the area of each triangle or parallelogram.

7.

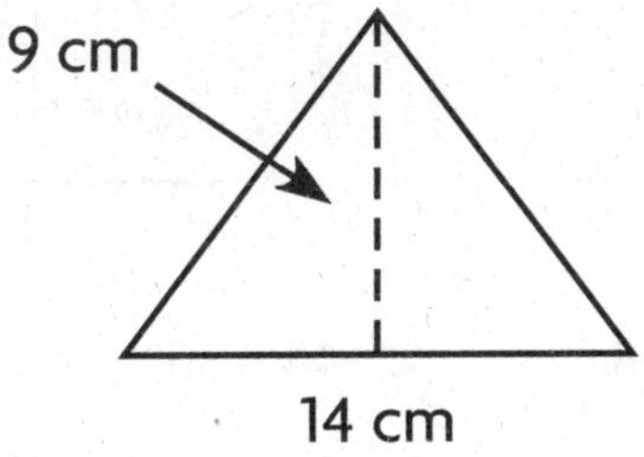

____

8.

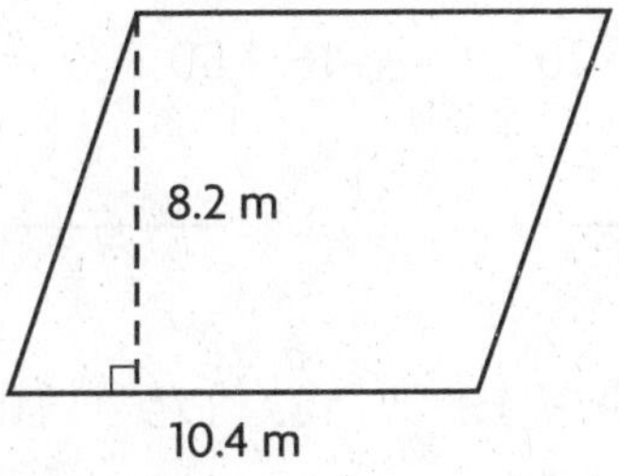

____

For 11–12, identify the faces that are parallel, perpendicular, or congruent. Name each solid figure that can be made with the net.

11.

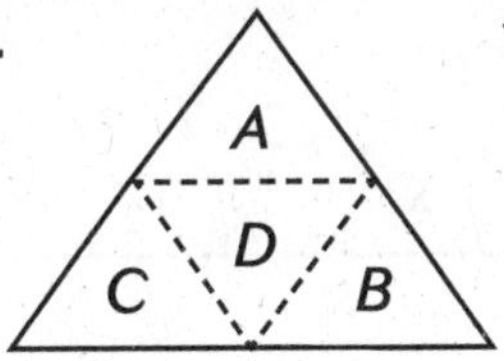

12.

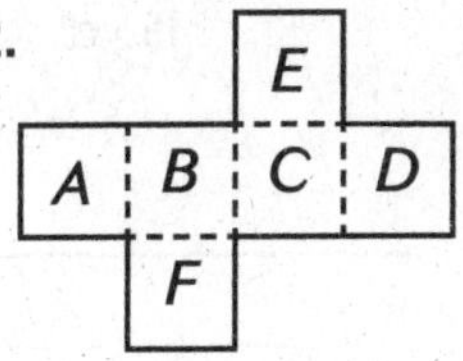

Name____ **Week 20**

Spiral Review

For 1–5, solve each problem.

1. What is the value of the underlined digit in $4{,}\underline{2}39{,}561$?

____

2. Write 2,345,587 in expanded form.

____

3. Write the standard form of three hundred three million, five hundred twenty-six thousand, ninety-one.

____

4. What digit is in the ten millions place in 25,080,795? ____________

5. Write 9,641,508 in word form.

____

____

____

For 8–10, use the double bar graph.

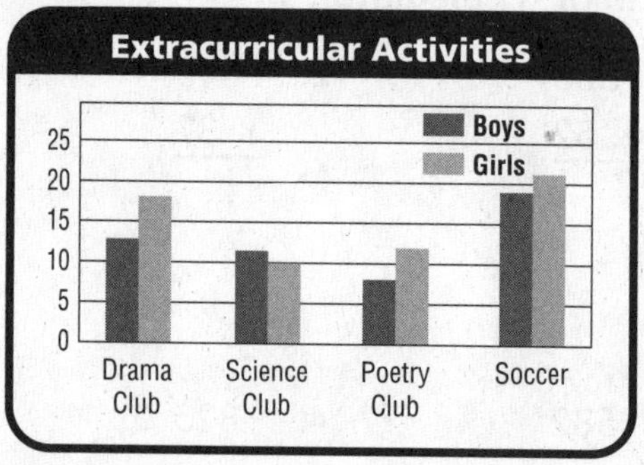

8. How many sets of data does the graph show? ____________

9. Which activity has the greatest number of girls? ____________

10. How many more girls than boys are signed up for drama club? ____________

For 6–7, find the volume of each rectangular prism.

6.

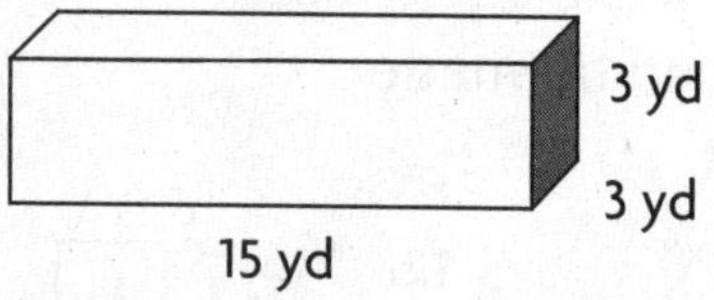

____

7.

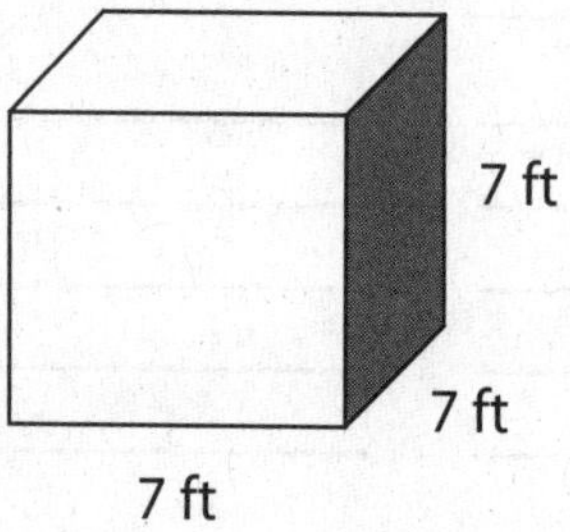

____

For 11–18, solve each equation. Check your solution.

11. $39 = 15 + r$ ____________

12. $3 \times n = 75$ ____________

13. $a \div 8 = 8$ ____________

14. $36 - w = 20$ ____________

15. $\$4 \times y = \$20$ ____________

16. $80 \div h = 4$ ____________

17. $y - 3 = 49 - 13$ ____________

18. $17b = 102$ ____________

Name________________________________ **Week 21**

Spiral Review

For 1–4, use basic facts and patterns to find the missing product or quotient.

1. 60 ÷ 10 ________
2. 630 ÷ 70 ________
3. 7,200 ÷ 8 ________
4. 4,800 ÷ 60 ________

For 5–8, divide. Check your answer.

5. $24\overline{)318}$
6. $72\overline{)609}$
7. $43\overline{)825}$
8. $67\overline{)936}$

A food company wants to know if people ages 18–40 like their new pasta. For 19–21, tell whether each sample represents the population. If it does not, explain.

19. a random sample of 500 women, ages 18–40

____

____

20. a random sample of 500 people, ages 18–40

____

____

21. a random sample of 500 adults

____

____

For 9–12, tell the appropriate unit for measuring each. Write *linear*, *square*, or *cubic*.

9. length ________
10. area ________
11. perimeter ________
12. volume ________

For 13–18, tell the units you would use for measuring each. Write *linear*, *square*, or *cubic*.

13. space inside a box ________
14. fence to enclose a yard ________
15. size of a classroom ________
16. wrapping paper for a box ________
17. distance around a package ________
18. width of a student desk ________

Graph a triangle with the given vertices. Then translate each vertex of the triangle 3 units down and 3 units left. Sketch and label the new triangle.

22. (4,5), (6,5), and (4,7)

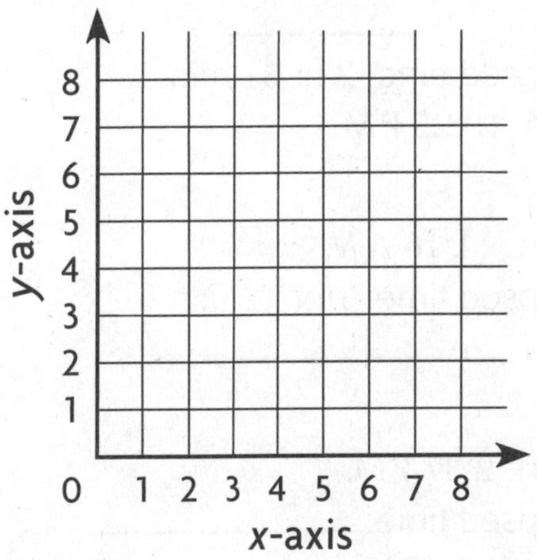

Spiral Review

For 1–2, write a mixed number and a fraction for each.

1. 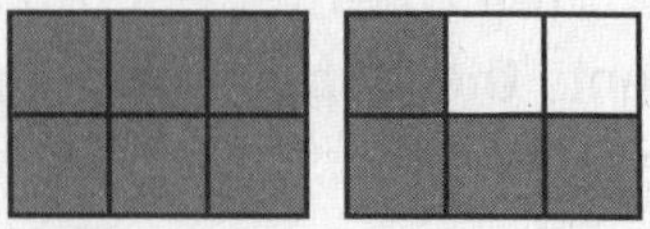 ________

2.  ________

For 3–4, identify each pair of numbers. Tell whether each pair is *equivalent* or *not equivalent*.

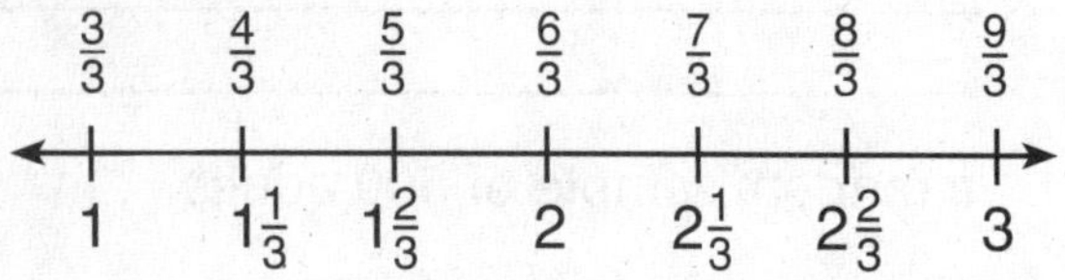

3. $1\frac{1}{3}$ and $\frac{5}{3}$ ________

4. $2\frac{2}{3}$ and $\frac{8}{9}$ ________

For 5–8, write the time for each.

5. start: 7:38 A.M.
   elapsed time: 3 hr 52 min
   end: ________

6. start: ________
   elapsed time: 2 hr 31 min
   end: 10:25 P.M.

7. start: 11:16 A.M.
   elapsed time: 1 hr 19 min
   end: ________

8. start: 2:37 P.M.
   elapsed time: ________
   end: 4:19 P.M.

For 9–12, use the table.

Number of Laps Run	
Student	**Laps**
Jill	6
Kelly	4
Rob	9
Mike	7
Sam	9

9. What is the order of the data arranged from least to greatest? ________

10. What is the median? ________

11. What is the mode? ________

12. What is the range? ________

For 13–15, find a rule. Write the rule as an equation.

13.

x	0	1	2	3	4
y	0	6	12	18	24

____

14.

x	12	10	8	6	4
y	6	5	4	3	2

____

15.

x	13	11	9	7	5
y	9	7	5	3	1

____

Spiral Review

For 1–3, compare. Write <, >, or = for each ○.

1. 0.754 ○ 0.734
2. 1.09 ○ 1.10
3. 10 ○ 0.909

For 4–7, order from greatest to least.

4. 1.345; 1.305; 1.354 ______________
5. 0.101; 0.110; 0.100 ______________
6. 73.806; 7.386; 73.860 ______________
7. 0.385; 0.853; 0.835 ______________

For 10–12, use the table. The table shows the results of a marble experiment.

Marble Experiment			
	Red	Blue	Green
Number of Pulls	𝍸 III	III	𝍸 IIII
Total	8	3	9

10. What is the experimental probability of pulling a red marble? ______________
11. What is the experimental probability of pulling a blue marble? ______________
12. What is the experimental probability of pulling a green marble? ______________

For 8–9, estimate the area of the shaded figure. Each square on the grid is 1 cm^2.

8. 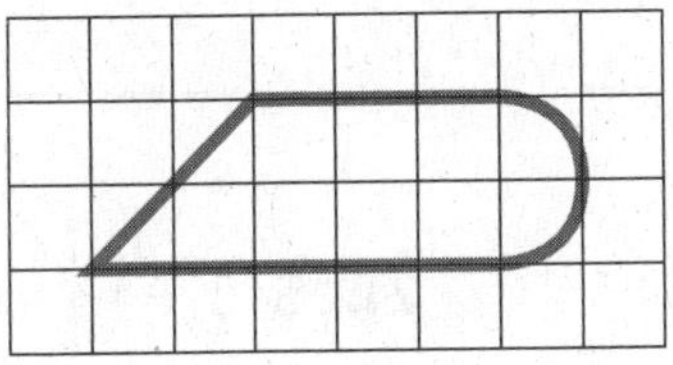

____

9. 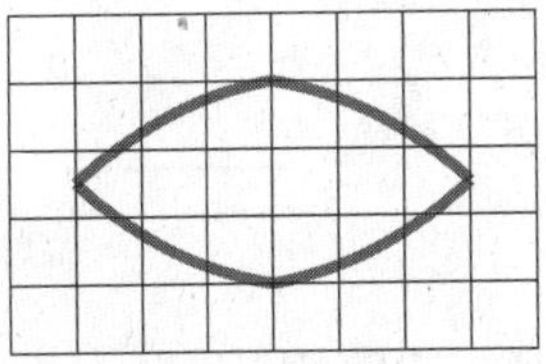

____

For 13–16, classify each figure in as many ways as possible. Write *quadrilateral, parallelogram, square, rectangle, rhombus,* or *trapezoid.*

13. 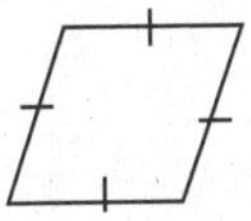

____
____

14. 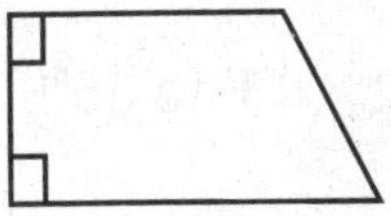

____
____

15. 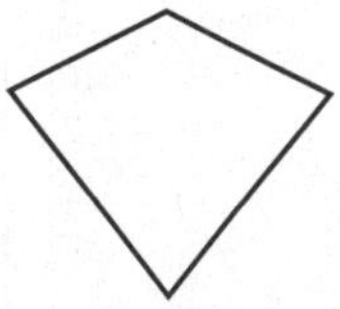

____
____

16. 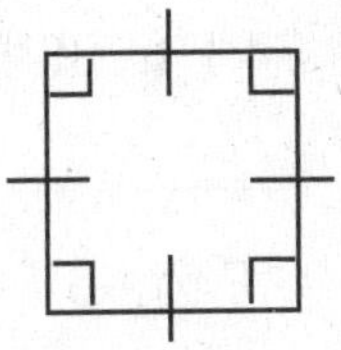

____
____

Name______________________________ **Week 24**

Spiral Review

For 1–10, find the sum or difference. Write it in simplest form.

1. $\frac{2}{5} + \frac{2}{5} =$ ____ 2. $\frac{1}{8} + \frac{3}{8} =$ ____

3. $\frac{4}{9} - \frac{1}{9} =$ ____ 4. $\frac{5}{7} - \frac{2}{7} =$ ____

5. $\frac{4}{12} + \frac{6}{12} =$ ____ 6. $\frac{3}{4} - \frac{1}{4} =$ ____

7. $\frac{6}{10} + \frac{2}{10} =$ ____ 8. $\frac{8}{9} - \frac{2}{9} =$ ____

9. $\frac{5}{6} - \frac{2}{6} =$ ____ 10. $\frac{1}{3} + \frac{1}{3} =$ ____

For 16–17, use the table to find the experimental probability. Then predict the outcome of future trials.

16. number of green tiles in 40 more pulls

Tile Pulls		
Green	Red	Orange
||	卌	|||

____

17. number of wins in 36 more games

Games	
Wins	Losses
卌 |||	||||

____

For 11–15, Find the change in temperature.

11. 12°F to 31°F ____________

12. 0°F to 35°F ____________

13. –10°F to 7°F ____________

14. 74°F to 88°F ____________

15. 0°F to –6°F ____________

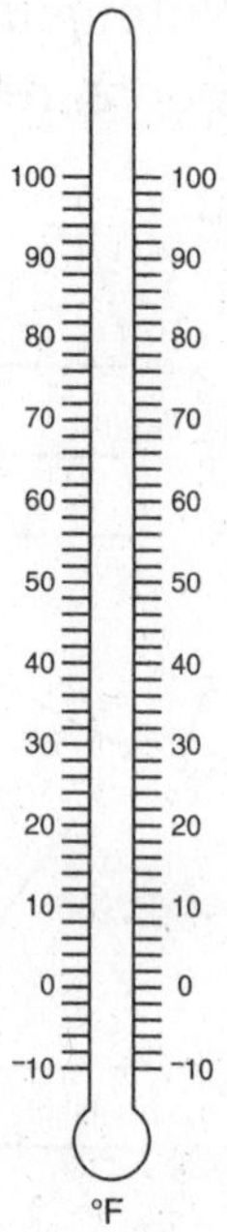

For 18–25, solve each equation. Check your solution.

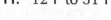

18. $49 = h - 17$ ____________ 19. $24 \div a = 8$ ____________

20. $9 \times n = 54$ ____________ 21. $\$42 - w = \$35$ ____________

22. $3 \times y = 42$ ____________ 23. $h \div 7 = 4$ ____________

24. $d - 9 = 21 \div 3$ ____________ 25. $34 + 8 = n - 10$ ____________

Spiral Review

For 1–6, compare.
Write <, >, or = for each ◯.

1. $\frac{3}{8}$ ◯ $\frac{2}{3}$

2. $\frac{3}{5}$ ◯ $\frac{4}{7}$

3. $3\frac{4}{9}$ ◯ $3\frac{2}{7}$

4. $5\frac{5}{7}$ ◯ $5\frac{4}{5}$

5. $3\frac{1}{2}$ ◯ $3\frac{5}{14}$

6. $1\frac{3}{4}$ ◯ $2\frac{1}{3}$

For 7–8, write in order from least to greatest.

7. $\frac{5}{8}, \frac{1}{4}, \frac{3}{8}$ ____________

8. $1\frac{5}{6}, 2\frac{3}{4}, 1\frac{1}{2}$ ____________

For 11–13, use the graph.

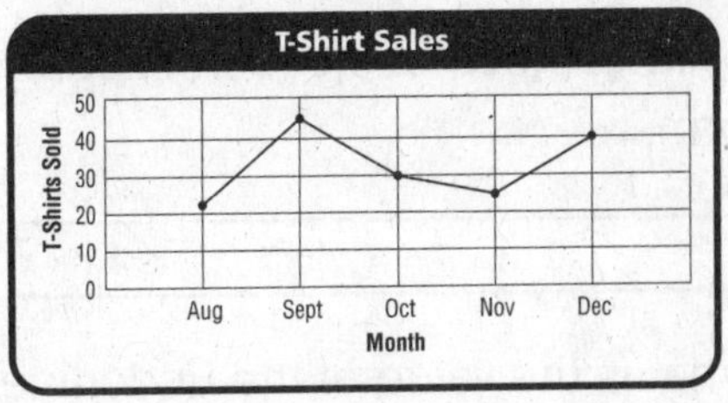

11. What scale and interval are used in the line graph?

____

12. Write the points on the line graph as ordered pairs.

____

13. What is the median? How would the graph change if the median sale were the same as September's sale?

____

For 9–10, find the area of each parallelogram.

9.

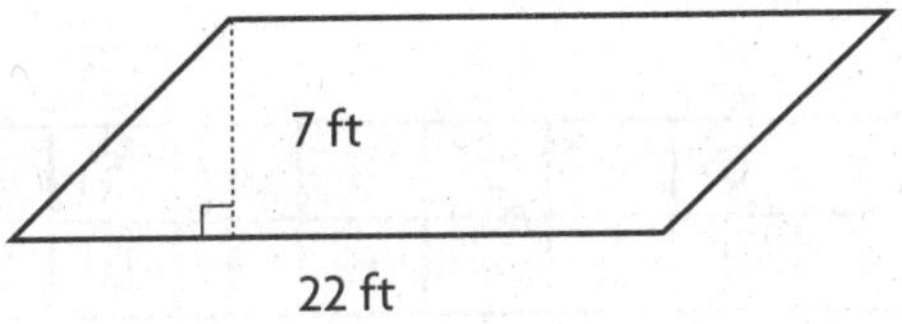

____

10.

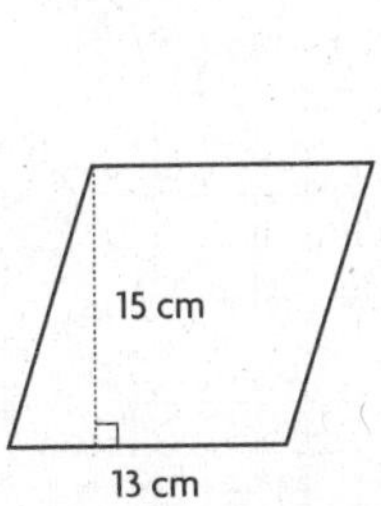

____

For 14–15, tell how the first figure was moved. Write *translation*, *reflection*, or *rotation*. For a rotation, write *clockwise* or *counterclockwise* and $\frac{1}{4}$, $\frac{1}{2}$, or $\frac{3}{4}$.

14.

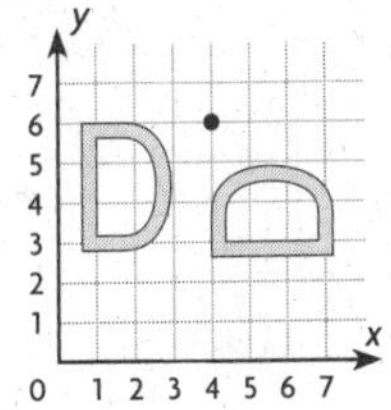

____

15.

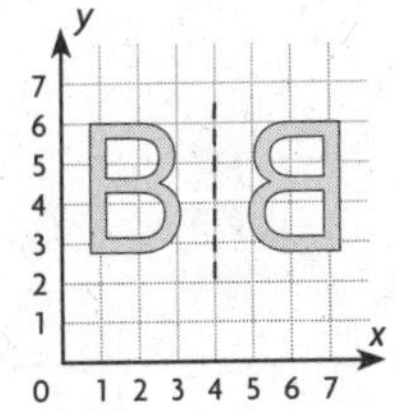

____

Spiral Review

For 1–5, solve each problem.

1. Write 690,303,520,002 in expanded form.

____

____

2. What is the value of the underlined digit in 32,405,922,287?

____

3. Write the standard form of five billion, six hundred ninety-six million, three hundred seventy-five thousand, twelve.

____

4. What digit is in the ten billions place in 670,050,213,604? ____________

5. Write the value of the underlined digit in 2,456,200,871.

____

Make a list or draw a tree diagram to find the total number of choices.

8. arrangement: ways to pull green, yellow, and blue tiles from a bag without looking

____

For 6–7, write whether you need to find perimeter, area, or volume to solve the problem. Then solve using the appropriate formula.

6. tile for this floor

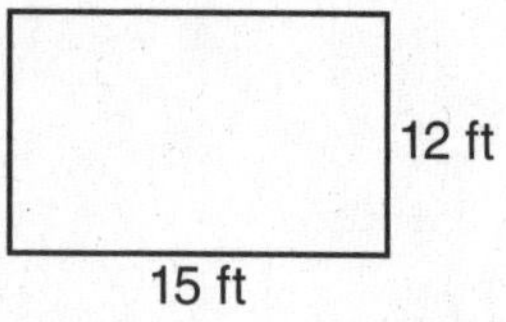

____

7. space inside this box

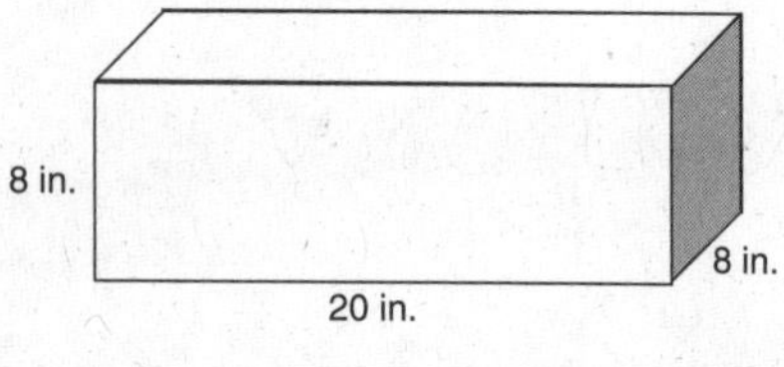

____

Write the rule as an equation to show the function. Then graph the ordered pairs.

9.

x	0	1	2	3	4	5	6
y	0	2	4	6	8	10	12

____

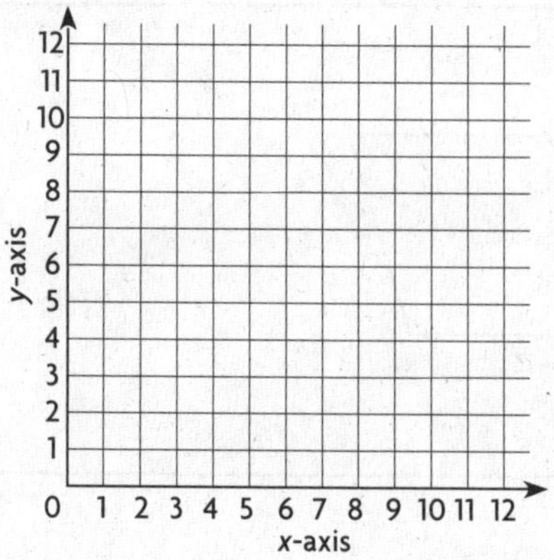

Spiral Review

For 1–6, find the sum or difference.

1. $85.19 + 37.48$

2. $251.895 - 75.362$

3. $7.081 + 6.254$

4. $3.582 - 0.763$

5. $0.85 + 0.063 + 3.572$

6. $11.804 + 6.137 + 15.749$

For 7–13, circle the unit you would use to measure each.

7. length of your hand
   inches or feet
8. length of a table
   miles or feet
9. length of an insect
   millimeters or meters
10. distance from New York to Michigan
    meters or kilometers
11. length of a soccer field
    inches or yards
12. height of a flag pole
    meters or kilometers
13. length of the classroom
    feet or yards

For 14–16, use the tally table.

Books Students Read		
Books	**Students**	**Frequency**
2	𝍸 |||	
3	𝍸 𝍸 |||	
4	|||	
5	𝍸 |	

14. Complete the frequency column in the table.

15. How many books read have the greatest frequency?

____

16. What is the range of the data?

____

For 17–24, write *acute, right,* or *obtuse* for each angle.

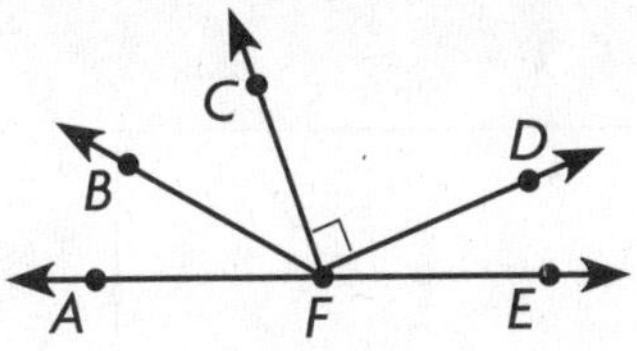

17. ∠*AFD* _______________
18. ∠*BFA* _______________
19. ∠*CFD* _______________
20. ∠*BFE* _______________
21. ∠*DFE* _______________
22. ∠*CFA* _______________
23. ∠*EFC* _______________
24. ∠*CFE* _______________

Name

Spiral Review

For 1–12, estimate the product.

1. 23×44
2. 61×28
3. 57×214
4. 46×697
5. 425×19
6. $\$7.68 \times 86$
7. 61×926
8. 584×73
9. 86×597
10. 243×36
11. $62 \times \$9.71$
12. 46×362

For 15–21, use the bag of marbles to write a fraction for the probability of the event of pulling the marble described.

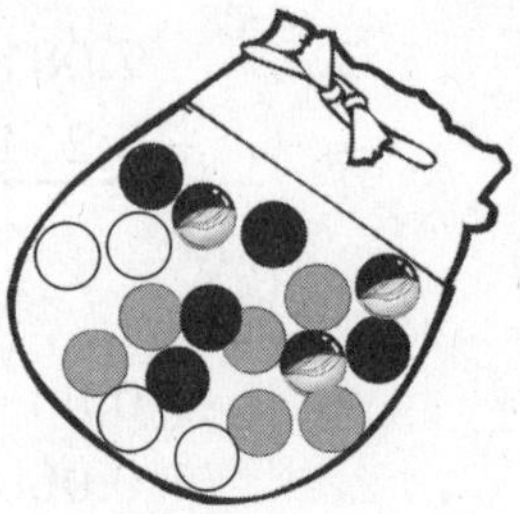

15. gray
16. black
17. white
18. swirl
19. gray or white
20. black, white, or gray
21. black or white

For 13–14, find the area of each figure.

13.

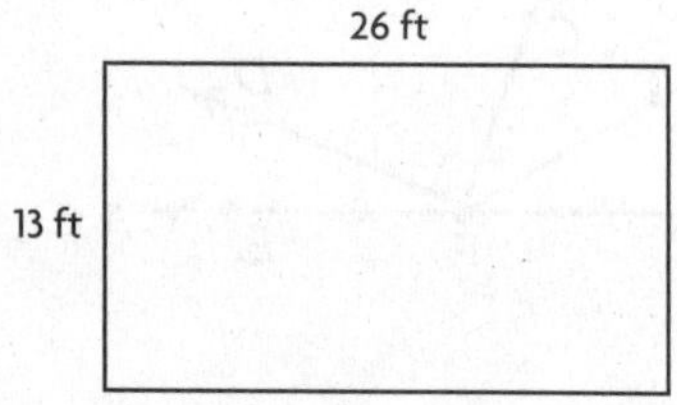

14.

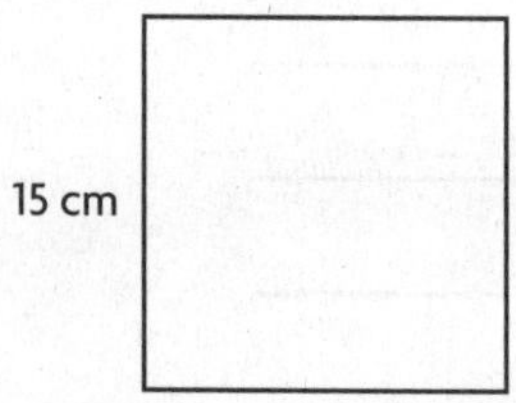

For 22–25, write a numerical expression. Tell what the expression represents.

22. Phillip bought 6 oranges at 99 cents each.
23. Hannah picked up 27 sticks in her front yard and 35 sticks in her back yard.
24. Tom had 14 gallons of gas in his car and then used up 8 gallons driving to work.
25. Lilly divided 24 grapes equally into 3 cups.

Spiral Review

For 1–8, estimate by rounding.

1. 29.63 + 18.05 ______

2. 87.905 − 38.714 ______

3. 4.139 + 7.652 ______

4. 2.763 − 0.509 ______

5. 93.47 − 62.13 __________

6. 11.042 + 8.765 __________

7. 43.869 − 10.062 __________

8. 0.654 − 0.398 __________

For 13–16, choose the best type of graph or plot for the data. Explain your choices.

13. number of students in 7 classrooms

____

____

14. hours people spend fishing

____

____

15. different seating sections of a stadium

____

____

16. population of deer over a 6-year period

____

____

For 9–12, write the time for each.

9. Start: 9:47 A.M.
Elapsed time: 2 hr 43 min
End: __________

10. Start: __________
Elapsed time: 3 hr 25 min
End: 8:16 P.M.

11. Start: 10:29 A.M.
Elapsed time: 2 hr 19 min
End: __________

12. Start: 3:15 P.M.
Elapsed time: __________
End: 4:57 P.M.

For 17–22, graph and label the following points on the coordinate grid

17. *A* (4,0) **18.** *B* (3,4) **19.** *C* (2,6)

20. *D* (8,1) **21.** *E* (5,2) **22.** *F* (1, 7)

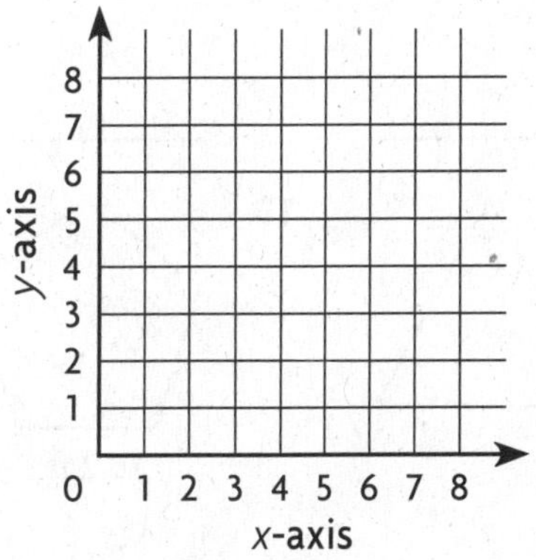

Name________________________________ **Week 30**

Spiral Review

For 1–6, estimate. Then find the product.

1. $\begin{array}{r} 315 \\ \times\ 57 \\ \hline \end{array}$

2. $\begin{array}{r} 642 \\ \times\ 38 \\ \hline \end{array}$

3. $\begin{array}{r} 493 \\ \times\ 62 \\ \hline \end{array}$

4. $\begin{array}{r} 510 \\ \times\ 26 \\ \hline \end{array}$

5. 37×628

6. 52×156

__________ __________

Draw a tree diagram to find the total number of outcomes.

11. a number cube labeled 1 to 6 and a coin tossed

____

For 7–10, find the perimeter of each polygon by using a formula.

7.
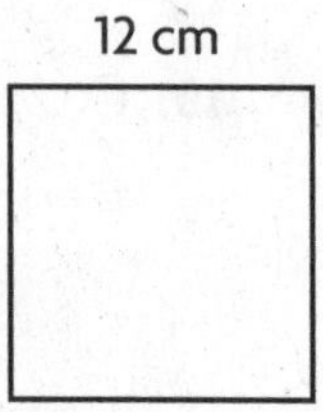

8.
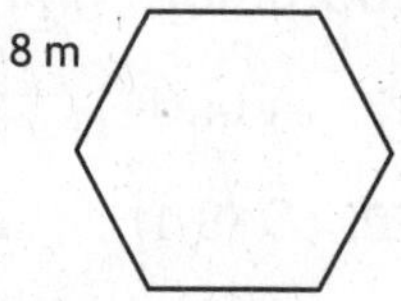

__________ __________

9.
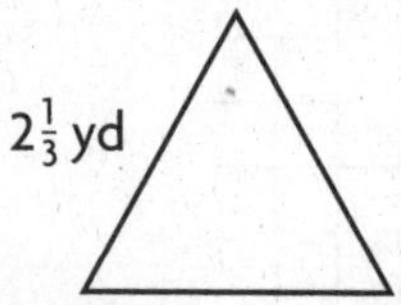

10.
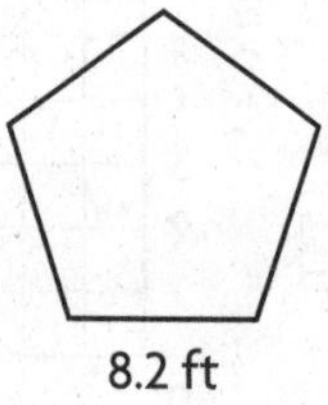

__________ __________

For 12–17, use counters to show all arrays for each number. Write *prime* or *composite*.

12. 7 __________

13. 27 __________

14. 16 __________

15. 81 __________

16. 3 __________

17. 12 __________

Name____________________________________ **Week 31**

Spiral Review

For 1–3, name the GCF of the numerator and denominator.

1. $\frac{8}{14}$ ____ 2. $\frac{12}{32}$ ____ 3. $\frac{12}{36}$ ____

For 4–9, write each fraction in simplest form.

4. $\frac{6}{15}$ ____ 5. $\frac{16}{28}$ ____ 6. $\frac{25}{40}$ ____

7. $\frac{14}{20}$ ____ 8. $\frac{30}{90}$ ____ 9. $\frac{48}{56}$ ____

For 10–11, complete.

10. $\frac{2}{3} = \frac{8}{\square}$ 11. $\frac{\square}{30} = \frac{1}{6}$

For 12–20, find the sum or difference.

12. 350 cm + 2.7 m = ____________

13. 15 m + 25 cm = ____________

14. 54 mm − 5.4 cm = ____________

15. 2.036 m − 36 mm = ____________

16. 3.45 km − 192.5 m = ____________

17. 6 ft 5 in. + 3 ft 6 in.

18. 12 yd 2 ft + 5 yd 1 ft

19. 9 ft 3 in. − 7 ft 4 in.

20. 12 yd − 3 yd 2 ft

For 21–23, use the line plot.

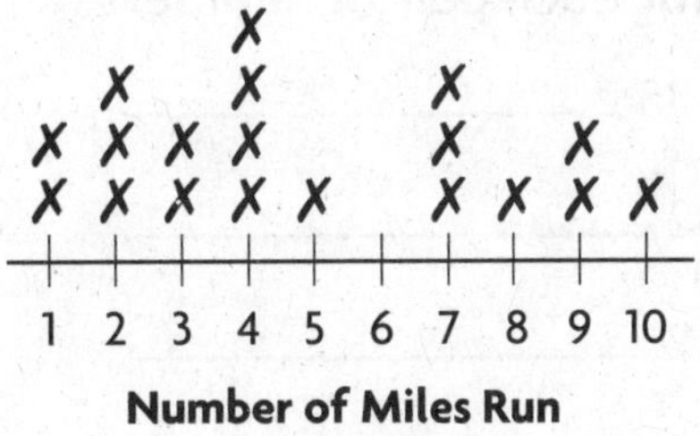

21. What is the median?

____

22. What is the mode?

____

23. What is the range?

____

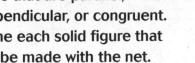

For 24–25, identify the faces that are parallel, perpendicular, or congruent. Name each solid figure that can be made with the net.

24.

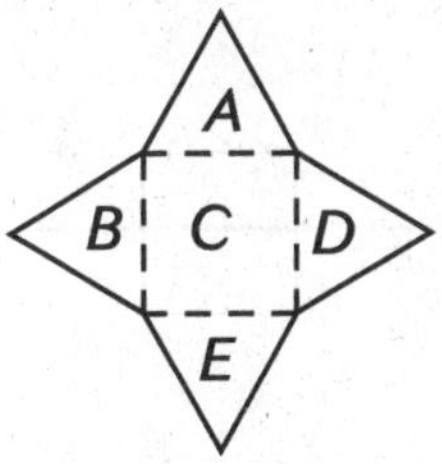

25.

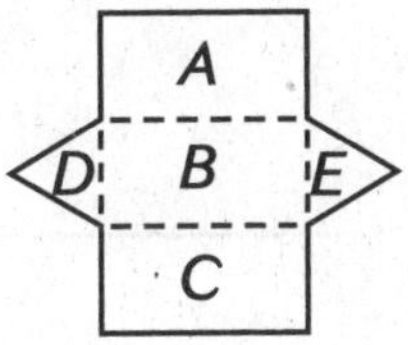

____________ ____________

____________ ____________

____________ ____________

____________ ____________

____________ ____________

Name

Week 32

Spiral Review

For 1–11, write the common factors for each pair of numbers.

1. 10, 35 ____________
2. 8, 32 ____________
3. 7, 42 ____________
4. 15, 45 ____________
5. 12, 30 ____________
6. 9, 27 ____________
7. 13, 26 ____________
8. 16, 40 ____________
9. 21, 63 ____________
10. 4, 20 ____________
11. 18, 24 ____________

For 14–16, express the experimental probability as a fraction in simplest form. Then predict the outcome of future trials.

14. 3 green sections in 18 spins. How many green sections in 24 more spins?

____

15. 6 red marbles out of 15 pulls How many red marbles in 35 more pulls?

____

16. 10 losses in 16 games How many losses in 40 more games?

____

For 12–13, find the volume of each rectangular prism.

12.

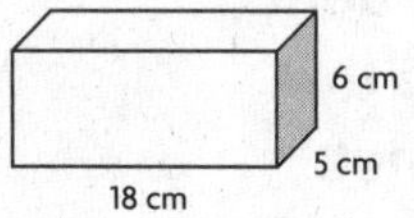

____

13.

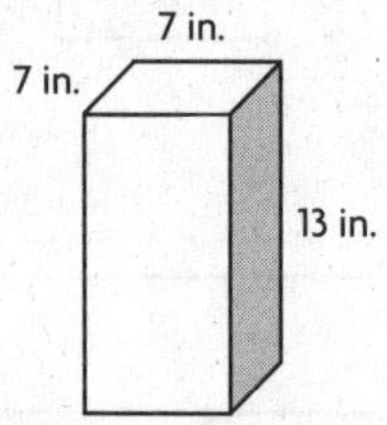

____

Write the rule as an equation to show the function. Then graph the ordered pairs.

17.

x	0	1	2	3	4
y	0	3	6	9	12

____

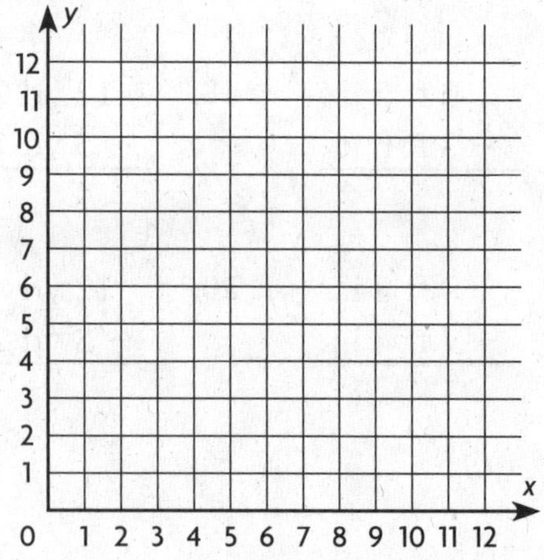

Spiral Review

For 1–4, write each mixed number as a fraction.

1. $1\frac{4}{5}$ ____ **2.** $2\frac{2}{3}$ ____

3. $1\frac{2}{7}$ ____ **4.** $3\frac{3}{8}$ ____

For 5– 8, write each fraction as a mixed number.

5. $\frac{8}{5}$ ____ **6.** $\frac{15}{13}$ ____

7. $\frac{17}{8}$ ____ **8.** $\frac{37}{12}$ ____

For 9–13, find the change in temperature.

9. 0°C to 18°C

____

10. −20°C to −5°C

____

11. −15°C to 10°C

____

12. 75°C to 10°C

____

13. 0°C to −16°C

____

For 14–16, use the graph.

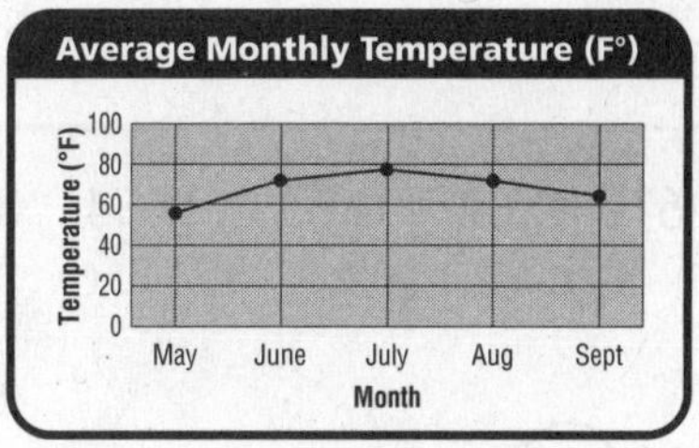

14. What scale and interval are used in the line graph?

____

15. How would you change the graph if the average daily temperature for August were 80° Fahrenheit?

____

16. What temperature is represented by the ordered pair (June, 73)?

____

Graph a triangle with the given vertices. Draw a horizontal line of reflection. Then reflect the triangle over the line. Sketch the new triangle.

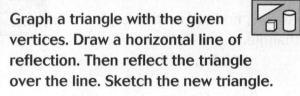

17. (2,7), (3,5), and (5, 6)

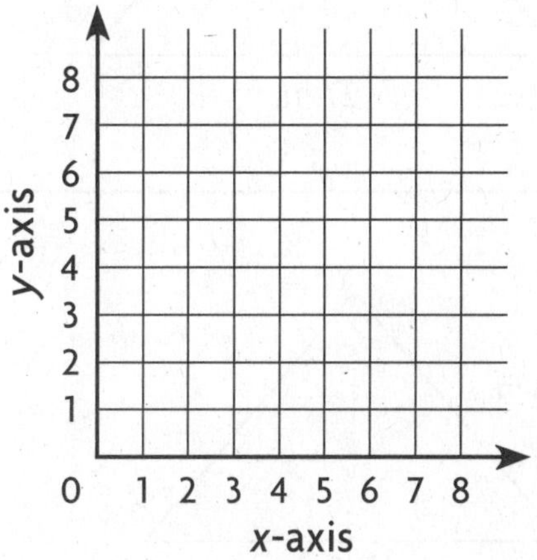

Spiral Review

For 1–6, write <, >, or = for each ◯.

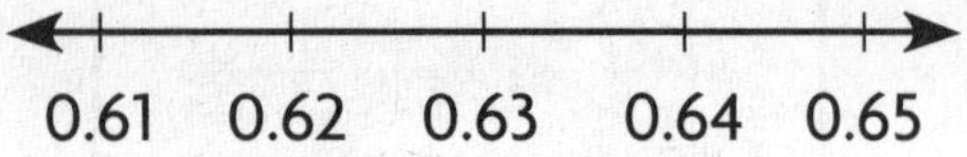

1. 0.643 ◯ 0.629
2. 0.617 ◯ 0.638
3. 0.649 ◯ 0.621
4. 0.640 ◯ 0.642
5. 0.620 ◯ 0.62
6. 0.637 ◯ 0.63

For 7–8, find the area of each triangle.

7.

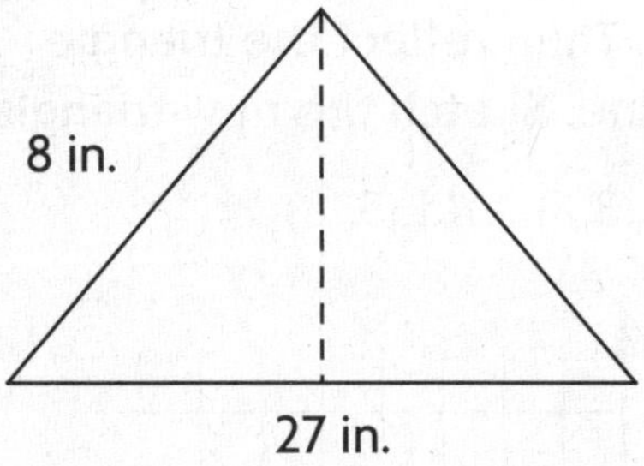

____

8.

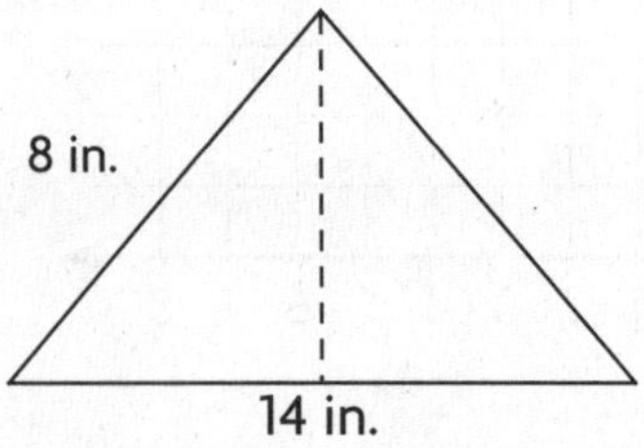

____

For 9–15, use a number cube labeled 1 to 6 to write a fraction for the probability of the event of tossing each number.

9. 3 ______
10. an odd number ______
11. a prime number ______
12. not 6 ______
13. a number greater than 4 ______
14. a number less than 8 ______
15. a multiple of 2 ______

For 16–18, write a numerical expression. Tell what the expression represents.

16. Kate had $30. Then she spent $8 to see a movie and $15 to buy a shirt.

____

____

17. Jim's 3 friends each found 7 shells.

____

____

18. Tyler scored 12 points in the first half and 17 points in the second half of a game.

____

____

Spiral Review

For 1–12, estimate the product.

1. 68 × 24 ________
2. 83 × 49 ________
3. 35 × 853 ________
4. 73 × 3985 ________
5. 568 × 31 ________
6. $8.28 × 76 ________
7. 34 × 964 ________
8. 672 × 95 ________
9. 42 × 754 ________
10. 529 × 51 ________
11. 74 × $8.36 ________
12. 54 × 762 ________

For 15–18, name the most appropriate graph for each.

15. Which type of graph would be most appropriate to show the heights of 23 basketball players?

____

16. Which type of graph would be most appropriate to record the growth of a plant over 5 weeks?

____

17. Which type of graph would be most appropriate to show the attendance for a week at the state fair?

____

18. Which type of graph would be most appropriate to show where a person's income is spent each month?

____

For 13–14, find the perimeter.

13.

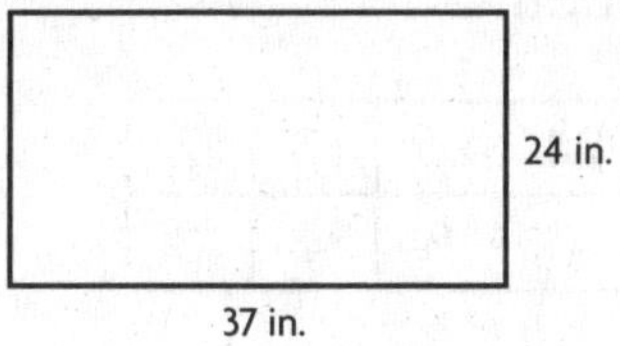

____

14.

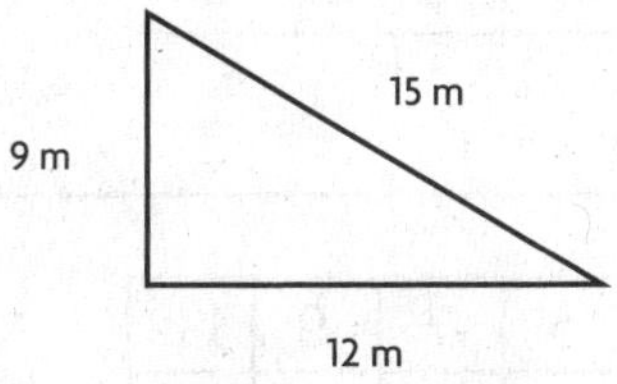

____

For 19–20, classify each triangle. Write *isosceles, scalene,* or *equilateral*.

19.

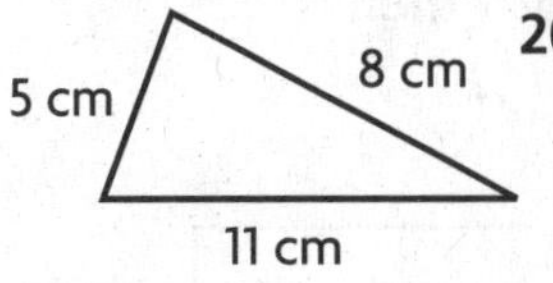

20.

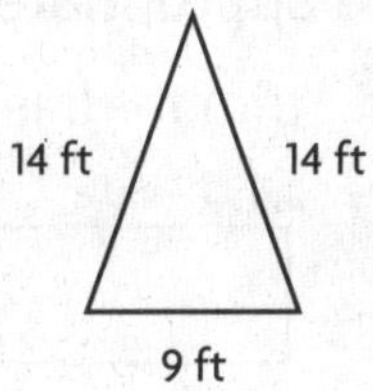

________ ________

For 21–22, classify each triangle. Write *acute, right* or *obtuse*.

21.

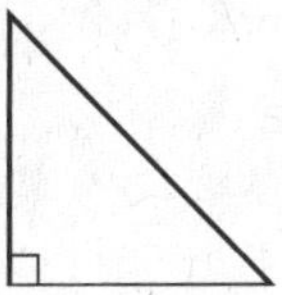

22.

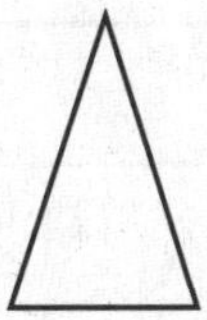

________ ________

Name_______________________________________ **Week 36**

Spiral Review

For 1–6, write each fraction as a decimal.

1. $\frac{4}{5}$ _____ 2. $\frac{7}{20}$ _____

3. $\frac{3}{10}$ _____ 4. $\frac{84}{100}$ _____

5. $\frac{35}{50}$ _____ 6. $\frac{78}{100}$ _____

For 7–12, write each decimal as a fraction in simplest form.

7. 0.2 _____ **8.** 0.38 _____ **9.** 0.57 _____

10. 0.46 _____ **11.** 0.65 _____ **12.** 0.44 _____

For 13–14, write whether you need to find perimeter, area, or volume to solve the problem. Then solve using the appropriate formula.

13. trim for this scarf

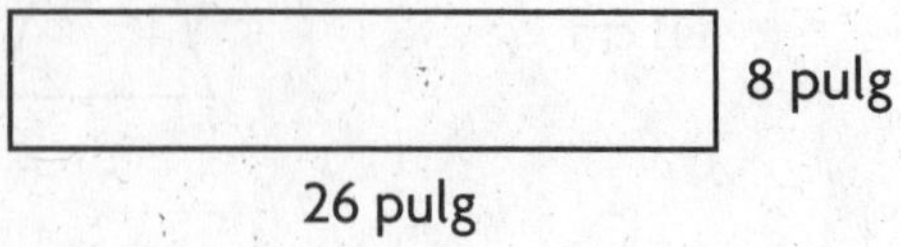

____

14. space in this storage box

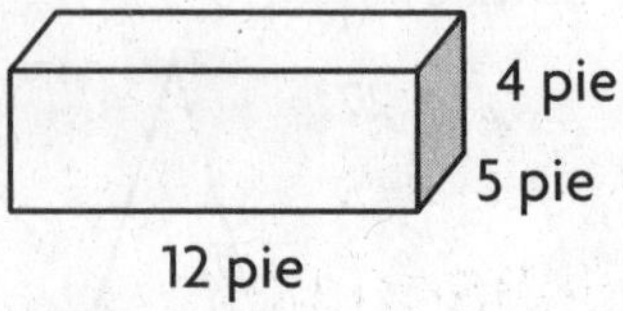

____

For 15–18, use the spinner. Write the probability of each event. Tell whether the event is certain, likely, unlikely, or impossible.

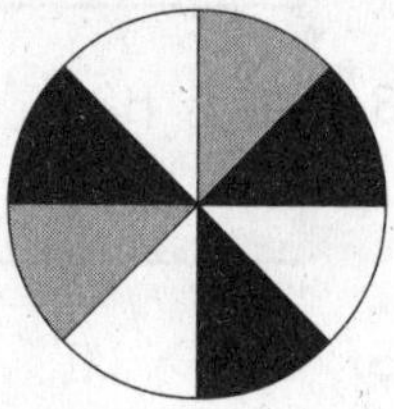

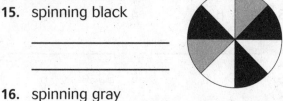

15. spinning black _______________ _______________

16. spinning gray _______________ _______________

17. spinning white or gray _______________ _______________

18. spinning green _______________ _______________

For 19–21, find a rule. Write the rule as an equation.

19.

input, x	24	20	16	12
output, y	6	5	4	3

____

20.

input, x	15	17	19	21
output, y	17	19	21	23

____

21.

input, x	5	7	9	11
output, y	35	49	63	77

____